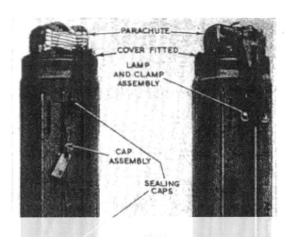

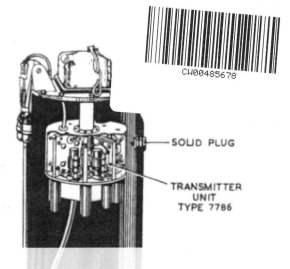

— SOLID PLUG

— TRANSMITTER
   UNIT
   TYPE 7786

# SONOBUOY HISTORY
# FROM A UK PERSPECTIVE

## RAE Farnborough's Role in Airborne Anti-Submarine Warfare

By Clive Radley –
Former Farnborough Project Manager

Foreword by Richard Gardner -
Aviation Writer, Farnborough Air Sciences Trust Chairman

**Fig. 1.  Transmitter Type T.7725—general views**

3.  In its pre-drop state all of the units of the sonobuoy are contained in a metal canister (fig. 1) suitable for fitting on a light series bomb carrier. The construction of the sonobuoy permits it to be dropped from an aircraft flying at a speed of up to 150 knots at a minimum altitude of 350 ft. or at speeds of up to 250 knots at altitudes between 500 ft. and 10000 ft.  When released from the aircraft, the rate of descent is controlled by a small parachute.

4.  On impact with the sea, the transmitter sub-assembly Type M1 (known as the lower or submerged unit) is released and slides out of the canister, sinking to a depth of 32 ft. determined by the length of the interconnecting cable.  The upper section of the cannister itself (the flotation chamber main assembly) is sealed and remains buoyant with a freeboard of about 4 in., suspending the sub-merged unit at the required depth.

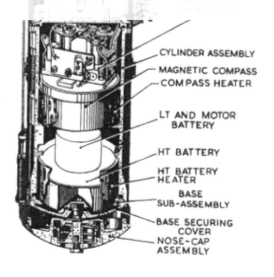

CYLINDER ASSEMBLY
MAGNETIC COMPASS
COMPASS HEATER

LT AND MOTOR
BATTERY

HT BATTERY

HT BATTERY
HEATER
BASE
SUB-ASSEMBLY

BASE SECURING
COVER
NOSE-CAP
ASSEMBLY

**Fig. 2.  Transmitter Type T.7725—sectional view**

# SONOBUOY HISTORY FROM A UK PERSPECTIVE

RAE Farnborough's Role in Cold War Anti- Submarine Warfare

The author has requested that the proceeds from the sale of this book should go to the following charity - The Farnborough Air Sciences Trust

Book design by Debbie Probert

Edited by Alan Wignall OBE, Mike Clapp, John Gilbert, Peter Evans, Les Ruskell and others

Scans of paintings provided by Hansen Fine Arts and Cranston Fine Arts

Front cover, specially commissioned painting, *Sonobuoy Sortie* - Lee Lacey associate member of the Guild of Aviation Artists

Back cover picture - Ultra Electronics Ltd

ISBN 978-0-9931694-2-7

# CONTENTS

# WHAT ARE SONOBUOYS

Sonobuoys are small expendable sonar systems which are dropped from anti-submarine warfare aircraft. The purpose of sonobuoys is to detect and locate the source of underwater sounds, primarily submarines. Sonobuoys are ejected from aircraft in canisters and after launch a parachute deploys to slow the descent. On impact with the sea it separates into two parts connected by a cable. An inflatable surface float with a radio transmitter rises to the surface for communication with the aircraft, while hydrophone(s) sensors and stabilising equipment descend below the surface to a preset depth set by the length of the cable. The sonobuoy radio transmits acoustic information from its hydrophone(s) to receiving equipment on the aircraft. The acoustic information is then processed using sophisticated algorithms and presented to operators on board the aircraft in both aural and visual form.

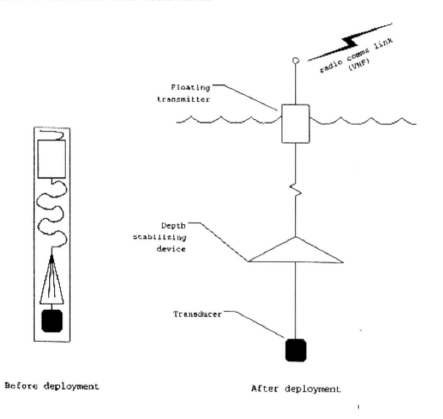

There are two main types of detection sonobuoy:

- Active - emit sound waves (pings) into the water and listen for returning echoes.

- Passive - emit nothing into the water but listen for mechanically generated sound (for instance, power-plant, propeller or door-closing and other noises) from ships or submarines. Can also listen for returning echoes when deployed in a field with active transmit sonobuoys.

Sonobuoys are capable of detecting surface ships, wave noise, marine life, oil rigs and in the case of active, wrecks and rocks as well as submarines. The acoustic processing on the aircraft aims to help the operator to rapidly distinguish between these nonsub returns and the submarine they are searching for.

In current sonobuoy systems the aircraft operator can transmit commands to the sonobuoy e.g. ping five times at intervals of x seconds.

To conduct a search for a submerged submarine a number of sonobuoys are usually deployed as a search field covering a predefined area the size of which depends on the type of target and the local sea state and sea bottom characteristics.

Sonar arrays on ships and submarines both hull mounted and towed can cost many millions of pounds as compared to an expendable passive sonobuoy where the cost can be hundreds of pounds and active transmit sonobuoys in the low thousands. However, search sonobuoys are deployed in multibuoy fields and as they are expendable each new search requires a new field of sonobuoys to be deployed.

The book adopts an overall total system approach to airborne ASW and accordingly the historical development of the UK's Maritime Patrol aircraft and their ASW relevant systems are described as well as the sonobuoys themselves.

# FOREWORD

## BY RICHARD GARDNER MCIPR MRAeS

This new book brings forward for the first time in such detail fascinating new information on the story of British Airborne Anti-Submarine Warfare development, and how sonobuoys transformed ASW capabilities, to the extent that the UK became a world leader in this aspect of modern warfare.

The author's close involvement at Farnborough with airborne ASW and his own meticulous research throws new light on how important this work was during World War Two and then even more so during the Cold War era. The results of all this effort were to be seen in the UK's outstanding contribution to NATO ASW operations, with systems deployable from Sea King and Nimrod aircraft proved themselves in all major air and sea exercises. The famous Fincastle Trophy annual ASW exercises, which ranged from US and Canadian waters to the seas around Australia and New Zealand, were won, time after time, by UK Nimrod crews, underlining the world lead that their MR1 and Mk2P platforms held over their equivalent US, NATO and Commonwealth equivalents, flying P-3s and Atlantiques.

The UK aerospace and defence industries, working alongside government R&D establishments, have also played a leading role in developing and delivering world-class ASW solutions since the submarine threat first arose.

The role of the RAE, later DRA and DERA (and Dstl), together with industry, has been very significant in developing systems that for decades gave the UK a highly effective counter to the perceived Russian submarine threat. Today that threat remains, and is once again growing, not diminishing, and many other nations, including China, are expanding their strategic nuclear submarine fleets. With the recent government U-turn on funding a new generation MPA, the task of developing suitable ASW systems to counter ever stealthier underwater vessels is likely to expand accordingly over the coming years. New opportunities will emerge and the need to continue R&D to stay ahead of the threat will be as important as ever

Richard Gardner MCIPR MRAeS
Farnborough, Hants
April 2016

# AUTHOR'S NOTE

My aim in writing this book was three-fold:

- To publicise and describe the contribution the RAE and its later incarnations made to the development of the UK's leading role in sonobuoy based anti-submarine warfare post world war 2 until the retirement of the Nimrod MR2 Maritime Patrol Aircraft in March 2010.

- To publicise the fact that sonobuoys are a major success story for UK industry with Ultra Electronics having become the world's leading supplier of sonobuoys.

- To raise money for the Farnborough Air Sciences Trust, (FAST), founded in 1993 to safeguard Farnborough's priceless aeronautical heritage.  FAST works to preserve and promote Farnborough's great heritage in aviation science - an important educational resource and an internationally significant part of the nation's scientific progress.

How I came to decide to write the book is rather convoluted, but I will try to explain. I had worked for most of my career up to 1994 at a company called EASAMS in defence systems consultancy in the Anti-Submarine Warfare domain. For a time, I had top secret clearance and was involved in OR (Operational Research) studies on contracts with the MOD, broadly aimed at helping to ensure that our deterrent submarines, then the Resolution class SSBN, could never be detected by USSR submarines and ASW forces. I visited the Clyde Submarine base at Faslane on a number of occasions and met the late Sandy Woodward when he was involved with submarine planning before he became famous during the Falklands War.

After the USSR collapsed and the Cold War ended MOD spending on ASW dropped and I was made redundant in late 1994 by EASAMS at the age of 49 when I was a Business Development manager. In early 1995 was then lucky to get another job in ASW not far from where I lived at the Defence Research Agency(DRA), formerly the RAE at Farnborough, by persuading the interviewers that I still remembered my technical knowledge in ASW. The position was in the Maritime Systems Group whose work was Airborne ASW. I then became project manager of a contract with that part of the MOD responsible for the procurement of sonobuoys. The contract provided a team of top DRA scientists and engineers as MOD's trusted internal technical advisors for sonobuoy development.

I retired in 2004 and became a volunteer at the Farnborough Air Sciences Trust Museum. Because of my background I was asked to produce an exhibit for the museum in the form of a rolling PowerPoint presentation describing the history of the development of sonobuoys in the UK and the RAE's role. After producing the presentation, I dropped out of FAST and rejoined in the summer of 2015. By this time a new sonobuoy exhibit was being planned and I agreed to project manage this activity and to write a book, this book.

# THE FARNBOROUGH AIR SCIENCES TRUST

http://www.airsciences.org.uk/aviation-history.html

FAST (the Farnborough Air Sciences Trust) was founded in 1993 to safeguard Farnborough's priceless aeronautical heritage. It works to preserve and promote Farnborough's great heritage in aviation science - an important educational resource and an internationally significant part of the nation's scientific progress.

The RAE and the other former MOD research establishments and their contractors produced a multitude of equipment, artefacts, technical reports, wind tunnel models, sonobuoys etc. during their lifetime. FAST aims to preserve this material and most of it is stored in containers by FAST but some of the more important ones are displayed at the FAST museum for the general public to view.

Finlegh Gordon, a retired RAF squadron leader, and Laurence Peskett a freelance architectural designer (left and right respectively).

Fin worked on a flight simulator at a 'lodger' company within the RAE site, and had been made aware (by vibrations cracking the walls of his building) of extraordinary RAE testing facilities behind the normally closed doors of strange buildings facing his office. Laurence had many years of experience working with historic buildings.

FAST is a registered charity (Reg No 1040199) and a fully accredited national museum. It is entirely self-funded through the activities of its Trustees and supporting members' association (FASTA), supplemented by donations from the general public. Additional sponsorship and grants are actively sought, alongside fund-raising events, museum hosting and shop sales.

Richard Gardner is a specialist aviation writer and trustee and is currently FAST's chairman. His numerous contacts with the aerospace industry were invaluable in securing support, both financial and moral, presenting FAST's proposals for a museum to various bodies including at Parliamentary level and, two years later, reversing serious opposition from a sector within the aviation industry itself.

Richard Gardner

The Nimrod model is one of many in the museum aircraft model room. This particular aircraft has been preserved and may be viewed at the Yorkshire Air Museum.

FAST Museum Nimrod MR2P model *(Author's picture)*

Sonobuoys in FAST container store 2015 *(Author's picture)*

# ACKNOWLEDGEMENTS

I would like to thank all those who have helped me in writing this book and in particular ex colleagues from the RAE, DRA, DERA, QinetiQ:

The ex-colleagues are:

Mike Clapp - currently an independent Maritime Systems consultant who supported DRA/DERA/QinetiQ on sonobuoy and related activities from 1995 to 2015 and who provided a major input to the Multistatics chapter, content advice and editing. Formerly a RN Naval Officer and business manager at Westland Systems Assessment.

The late John Davey - RAE/DRA/DERA/QinetiQ. Compiled and documented a sonobuoy history research paper.

Les Ruskell - Currently a FASTA volunteer but formerly Passive Sonobuoy Technical Leader 1980s RAE/DRA before moving the CDA in the 1990s. He wrote large parts of chapters 6, 8 and 10.

Dr Tony Heathershaw – ex Technical Leader Environmental Impact Assessment DRA/DERA etc Winfrith and Southampton Oceanology. Contributed to chapter 11.

Peter Viveash - Passive Sonobuoy Technical Leader 1990s RAE/DRA/DERA/QINETIQ -Contributed to chapters 4, 6, 8, 9, and 10.

Ray Cyphus – Active Sonar Technical Leader 1980s/1990s RAE/DRA/DERA/DSTL – Contributed to chapters 10 and 11.

Derrick McNeir – Airborne Processing Technical Leader 1980s – 2016 RAE/DRA/DERA/QINETIQ- Contributed to chapters 10 and 11.

Peter Evans – Ex Technical Leader AUWE Portland and Winfrith Active Sonar and Sonobuoy Systems – Contributed to number of chapters and provided advice on content.

I would also like to thank people from the wider research and industry community for reviewing or contributing material, as follows

Alan Wignall - OBE, BSc(Eng), MSc, CEng, MIET, Technical Director, Ultra Electronics Sonar Systems.

Bhupendra D Mistry - Technical Authority (Sonobuoys), Ultra Electronics Sonar Systems.

Peter Verburgt – Currently Navmar Applied Sciences Corporation but formerly Navair Pax River.

R A Holler - Currently Navmar Applied Sciences Corporation but formerly Navair Pax River.

Richard Gardner/Brian Luff/Graham Rood – Prominent members of the Farnborough Air Sciences Trust.

Tony Blackman - author of Nimrod: Rise and Fall.

John Gilbert - CEng, MRAeS, BSc(Hons), Working career in the Stability & Control and Flying Qualities of Aircraft mostly in Southern California, but also in Canada and the UK.

## ORGANISATIONS

- The National Archives Kew
- RAE/DRA/DERA/QINETIQ
- The Farnborough Air Sciences Trust
- UK MOD
- Ultra Electronics Ltd Greenford
- The RAF Historical Society
- The Shackleton Association
- NASC/NAVMAR
- The Guild of Aviation Artists
- Hansen Fine Arts

In the last few weeks before going to print John Gilbert, and Alan Wignall spent an amazing amount of time copy editing and correcting errors etc. Many thanks.

A number of ex colleagues at EASAMS, DRA/DERA/QinetiQ and customers have influenced my career in a positive way such that without them this book probably wouldn't have been written.

They are:

EASAMS - Mike Dodd, John York, Dr Bernard Harvey, Steve Moore, Peter Wills, Brian Knapp, Tom Ferrier, Steve Crawford, Angus Cairns, John Hill, Geoff Curzon , Ken Brett, Len Martin.

DRA/DERA/QinetiQ - Neil Andrew, David Cooke, Ray Cyphus, Peter Viveash, Ben Wynne, Mary Simmonds, Peter Martinson, Mike Clapp, Mark Brown, Colin Snow, Andrew Probyn, Mike Nicholas, Avtar Gida, Roy Frost, Alex Crowe, Ken Clayton, David North, Lee Cooper, Julian Fletcher, Mel Martin, Trevor Kirby Smith, Steve Riley, Simon Vines

PPAG - Peter Crawley, Neville Wood, John Webber,

AUWE - Dennis Howell, Dr Malcolm Woollings

DOR(Sea) - Shane Heaney

Nimrod MRA4 IPT - John Gillett

Finally, I owe an enormous debt to Christopher Dodd who co-founded the River & Rowing Museum at Henley on Thames in 1994 after 30 years as an editor in the Guardians Features department, and is now a rowing historian. It was Chris who first encouraged me to become a self publishing author and this book is my second.

# INTRODUCTION

The concept of aircraft dropping acoustic sensors (sonobuoys) into the ocean for detecting submerged submarines from aircraft was the brainchild of the UK's Professor Patrick Maynard Stuart Blackett during World War 2 to help counter the U- Boat threat to Allied convoys.

Professor Patrick Maynard
Stuart Blackett
*Wikipedia*

Professor Blackett was Director of Operational Research with the UK Admiralty from 1942 to 1945 based in London. He was one of Churchill's Scientists as described in the winter 2015/6 exhibit at the Science Museum in London.

*'More than any other country in the world, Britain must depend for survival on skilled minds.'* So, said Winston Churchill when he opened Churchill College, Cambridge in October 1959. It summed up his life-long belief that, if Britain could gain any advantage over an adversary by drawing upon the ideas of its scientists and inventors, then they should be encouraged. He adopted this approach during world war 2 and that period generated an immense number of scientific leaps and technological advances.

Churchill immersed himself in the work of the nation's engineers, mathematicians and physicists and a new word appeared, 'boffin', to describe the scientists who worked away on things which most people did not even begin to understand. But Churchill knew that among them were probably the ideas that would help win the war and he wanted them realised as soon as possible.

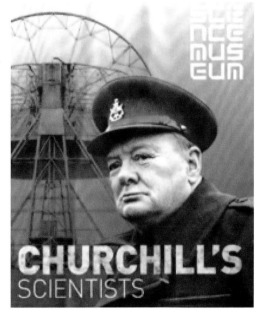

Professor Blackett was one of Churchill's 'boffins' and a pioneer of Defence Operational Research (OR) and later in World War 2 he wrote the following which illustrates his philosophy.

*New weapons for old is apt to become a very popular cry. The success of some new devices has led to a form of escapism, which runs somewhat thus: Our present equipment doesn't work very well, training is bad, supply is poor, spare parts non-existent. Let's have an entirely new gadget. Then comes the vision of a new gadget, springing like Aphrodite from the Ministry of Aircraft Production, and attended by a chorus of trained crews. One of the tasks of the OR section is to attempt to make a numerical estimate of the merits of a change over from one device to another, by continual investigation of the performance of existing weapons, and by objective analysis of the likely performance of new ones.*

The use of numerical OR inputs to aid defence procurement decision making was adopted by the UK, the USA and a number of other countries and continues in the 21 st century.

Professor Blackett's ideas led to the first operational sonobuoy being developed in the USA in world war 2 with assistance from the UK and was manufactured and mass produced there.

The U-Boat war during battle of the Atlantic has been described as the one that really did threaten Britain's survival as surely would have panzer divisions roaming through the Home Counties. In his memoirs Churchill wrote that THE ONLY THING THAT EVER REALLY FRIGHTENED ME during world war 2 was the U-Boat peril. He said he was more anxious about U-Boats than the outcome of the Battle of Britain.

# CHAPTER 1

## A SUMMARY HISTORY OF THE RAE
## AND ASSOCIATED MOD RESEARCH ORGANISATIONS

In 1904-06 the UK's Army Balloon Factory relocated from Aldershot to the edge of Farnborough Common in order to have enough space for experimental work. In October 1908 Samuel Cody made the first aeroplane flight in Britain at Farnborough and in 1912 the Balloon Factory was renamed the Royal Aircraft Factory (RAF). In 1918 the Royal Aircraft Factory was again renamed, becoming the Royal Aircraft Establishment (RAE) to avoid confusion with the Royal Air Force, which was formed on 1 April 1918.

**The Royal Aircraft Establishment (RAE)** at Farnborough then became well known as a world leading centre for aerospace research and development In 1988 the RAE was renamed the Royal Aerospace Establishment and in 1991 this became part of the Defence Research Agency (DRA), the MOD's new research organisation which included the Admiralty Underwater Weapons Establishment, AUWE, at Portland.

Samuel Cody Statue Unveiling Ceremony at Farnborough *(Author's Picture)*

After World War 2 to further research carried out at Portland for the Admiralty, a new Admiralty Gunnery Establishment was built at Portland. The work commenced in 1949 and was completed during 1952. In 1959 the AGE site became part of the Admiralty Underwater Weapons Establishment (AUWE). This amalgamation meant that MOD research into underwater weapons was moved to Portland, including work with the highest security classification at the height of the Cold War. Throughout its working life, the Southwell establishment worked alongside the East Weares establishment, and both became responsible for the design, development and testing of underwater weapon and detection systems.

His Majesties Under Water Detection Establishment (HMUDE) was another Portland based MOD research establishment which as its name implied, was all to do with underwater detection, and developed hydrophones and underwater sensors. It was originally part of HMS Osprey at Portland Naval base (AUWE North) and came into existence as a result of the U-Boat problems of World War 1. HMUDE became part of AUWE in the 1960s to form AUWE.

In 1961 Portland's two establishments at Southwell and East Weares were the centre of worldwide attention, after the discovery of espionage infiltration. This became infamously known as the Portland Spy Ring, a Soviet spy ring that operated in England from the late 1950s until 1961 when the core of the network was arrested by the British security services. In 1984 the AUWE became part of the Admiralty Research Agency (ARE) when all naval research came under the same direction and this in turn became part of the joint service Defence Research Agency (DRA) in 1991. At the end of the Cold War, an announcement was made that both the navy base and the ARE research establishments were to close in 1995. The buildings were left empty, until the site was sold in 1997, and became the Southwell Business Park.

In 1995, the DRA and other MOD Research organisations merged to form the Defence Evaluation and Research Agency (DERA), and QinetiQ was formed in July 2001, when the Ministry of Defence (MOD) split its Defence Evaluation and Research Agency (DERA) in two.

The smaller portion of DERA, was rebranded Dstl (Defence Science & Technology Laboratory) and remained part of the MOD. The larger part of DERA, including most of the non-nuclear testing and evaluation establishments, was renamed QinetiQ and prepared for privatisation.

# CHAPTER 2

## THE HISTORICAL BACKGROUND WHICH LED TO THE DEVELOPMENT OF AIR DEPLOYED SONOBUOYS IN WORLD WAR 2

In both World Wars 1 and 2, German submarines posed a great threat to Allied surface ships. In World War 1 there was no real ASW (anti-submarine warfare) technology. The frigates and destroyers caught the U boats on the surface. There were no acoustic based detection devices (sonar and sonobuoys) that are commonplace today. However, the idea of using aircraft as a means of detecting submarines had first been mooted in 1911/12.

Emile Aubrun in a Deperdussin
*(aviation.maisons-champagne.com)*

Emile Aubrun an early flying pioneer in October 1911 using a Deperdussin monoplane aircraft had visually detected a partially submerged submarine and a periscope of another submarine in an experiment in the Cherbourg area. Armand Deperdussin was a rich forward-thinking silk merchant who employed Louis Béchereau to design and produce aircraft for him.

Hugh Alexander Williamson
1885-1979 *(Wikipedia)*

In 1912 the potential value of air surveillance of shipping was put to the UK Admiralty by a young RN officer, Hugh Williamson, who was a submarine commander and had learnt to fly. He wrote a paper, The Aeroplane against Submarines, which described the then ground-breaking idea that the presence of an aircraft would result in hostile submarines submerging to avoid visual detection which would drastically restrict their mobility. He suggested that the visual detection range on a surfaced submarine from an aircraft would be significantly longer than the submarine's crew's visual counter detection range on the aircraft. He also described the desired characteristics of submarine surveillance aircraft and surveillance tactics. He suggested a monoplane with 5 hours' endurance with a passenger to act as observer endurance for search.

Not surprisingly there were no aircraft with the desired characteristics and their engines then were very unreliable, it was only a few years since the Wright Brothers' first powered flight. He also proposed anti-submarine bombs with two fuses the first of which would operate when the bomb hit the water to ensure they would not detonate prematurely on the aircraft.

The Admiralty were impressed, and the RN carried out some trials and discovered that the wake of a submarine periscope could be visually detected at a tactically useful range as could surfaced submarines.

By 1915 during World War 1 there were still no suitable aircraft for anti submarine surveillance so the RN procured a number of small non-rigid airships which came to be called blimps and very quickly went into operational service to conduct anti-submarine surveillance. The disadvantage of blimps was that they were detectable at quite long ranges by U-boat crews, 10 miles, whereas the blimp crew's detection range was around five miles.

Over 200 blimps had been delivered by the time the war ended and approximately 100 were still in service. They proved to be an effective deterrent to attacks as U-boats tended to dive to avoid visual detection, thus limiting their mobility. Kite balloons deployed from escorts at heights of 3000 ft were also used to deter U-boat attacks. In the last 18 months of the war, 257 merchantmen were sunk but of these only two were lost from convoys with an aerial escort.

The S.S. Non-Rigid Airship in service from 1915

Aircraft and engine technology advanced rapidly during the war and by 1916 flying boats with sufficient endurance to carry out visual submarine surveillance endurance were available, and many were built. By the end of World War 1 airborne anti-submarine patrols above and around convoys had become the norm. Sopwith floatplanes patrolled over the North Sea and in the Med during the Dardanelles campaign attempting to detect U boats visually. There is no evidence that they ever sunk a U boat but their presence would have forced any U boat to remain submerged.

The Curtiss Large America flying boat patrolled over the North Sea on ASW patrols in World War 1 but had a weak hull and was replaced by the UK's Felixstowe F.2A in 1917 which was based on the Large America but was more suited to North Sea conditions.

RNAS Curtiss Large America

Felixstowe F.2A    Hansen Fine Arts

Between the wars surface ship sonars were developed by the Allies known as ASDIC, primarily active in operation i.e. pings were transmitted and echoes from any close by submarines were received by the ASDIC's receiving hydrophones. Little effort was devoted by the UK before World War 2 to giving aircraft the ability to detect submarines by non-visual means as ASDIC was thought to be the answer before war broke out. Also, during the inter war years the RAF gave anti-submarine operations low priority and Coastal Command had to make do with obsolescent aircraft.

Because of their misplaced confidence in ASDIC the Admiralty forgot the lessons of World War 1 re the value of airborne surveillance. U boats during the early and middle part of World War 2 in the main shadowed and closed on convoys on the surface prior to attacking as their underwater endurance was very limited. Unfortunately, ASDIC had very short detection ranges against submarines on the surface and also in shallow water and a number of other operational situations. Another ASDIC problem was that non-sub echoes were also detected resulting in false alarms. During this period the Germans concentrated on underwater listening systems for her U boats to enable them to detect surface ships without the need for pinging with active sonar.

The early World War 2 U boats were in reality surface vessels in the way they operated which could spend short periods of time submerged. The text and picture below are from http://uboat.net/types/ixb.htm. The VIIC was a typical early World War 2 U boat.

The VIIC was the workhorse of the German U boat force in World War Two from 1941 onwards and boats of this type were being built throughout the war. The first VIIC boat being commissioned was the U-69 in 1940. The VIIC was an effective fighting machine and was seen in almost all areas where the U-boat force operated although their range was not as great as the one of the larger IX types.

The VIIC came into service as the "Happy Days" were almost over and it was this boat that faced the final defeat to the Allied anti-submarine campaign in late 1943 and 1944.

Perhaps the most famous VIIC boat was the U-96 which is featured in the film Das Boot...

U-760 of type VIIC

The development of radar by the UK for detecting German bombers was just in time for World War 2 but it was not until 1941 that a crude radar was fitted in aircraft. Radar provided a useful tool for detecting surfaced U-boats and also had the effect of forcing them to dive and sometimes prevented them closing for attack on convoys. However, aircraft still lacked a means of detecting submerged U-boats.

During the early 1940s convoys supplying Britain were being attacked by U-boats operating in "Wolfpacks", which would carry out attacks from astern of the convoy and shipping losses were large.

Wolfpacks, were created by Karl Doenitz as a means to defeat the allied convoy system after his experiences as U-boat commander in World War 1. In June 1940 the first such operations were tried with the tactical control given to the senior officer of the group. The idea is simple enough; gather U-boats in patrol lines to scout for convoys. Once a convoy was spotted the first boat was designated "shadower" and would chase the convoy and report its heading and speed to the BdU, the Command HQ of the U-boat arm. The BdU knew of the daily positions of the U-boats and coordinated the operation against the convoy by ordering the nearby boats to form up around a convoy and to attack with as many boats as possible during the same night to overwhelm the escorts.

During 1941 1118 allied ships were sunk as compared to 30 U boats destroyed.

To help counter the U boat threat the UK set up a special UK Admiralty Committee, tasked with proposing anti-submarine measures, drawing on Operational Research techniques. Professor P M S Blackett, an eminent UK scientist was Director of this committee and first put forward the idea of an expendable sonar system, or sonobuoy as they later became known, in May 1941. He proposed that buoys equipped with hydrophones be dropped from suitable aircraft to listen for submarines attempting to attack a convoy. These special buoys would be fitted with radio transmitters, which would relay the hydrophone information to receivers in the aircraft. These devices would be called "Radio Sonobuoys."

**Left:** *The Consolidated Catalina, otherwise the PBY-5 of the US Navy, began to arrive in Coastal Command early in 1941. Slow but highly reliable, it could stay in the air for about 27 hours when fitted with extra fuel tanks. RAF Catalinas served in the North Atlantic, the Arctic, the Mediterranean, the Indian Ocean and the Far East. Detachments were also sent to North Russia to protect the arctic convoys. This photograph of a Catalina armed with four 250 lb depth charges was taken at moorings in Iceland.*

*Source: Aeroplane Monthly*

Thus, the concept of the sonobuoy was British. However, the British could not spare the technical resources to develop the idea at that time and the US led the initial development effort. RCA Camden and Columbia University at New London, Connecticut carried out initial development work in a joint war effort with the Sonar Research Unit Portland and the Royal Aircraft Establishment Farnborough.

Proof of concept sonobuoys were ship launched, weighed 60 lb (27kg), and were first tested in September 1941 and, despite teething problems, proved the concept. Typical problems where they tended to leak, sank and were noisy. Development of an airdropped version rapidly followed.

The need for such a passive detection device was highlighted by the fact that in 1941 95% of U boats detected by other means such as visual, radar or ASDIC escaped. The majority of these initial detections warned the U boat such that they dived and moved away.

It is estimated that after sonobuoys became operational about 40% of U boats destroyed were by aircraft or aircraft operating with surface ships.

By 1943, Germany had developed the snorkel which when fitted would enable U boats to expel engine exhaust fumes and take in fresh air at periscope depth and increased their underwater endurance. This made them far less detectable by radar. Fortunately for the Allies the snorkel equipped XXI class did not begin to enter service until March 1945 too late to influence the war. The XXIs were the first true submarines with a streamlined shape, much increased battery capacity, giving a tactically useful endurance on battery power alone. They were the last large U boats built by the Germans in World War 2. The following paragraph and picture is from http://uboat.net/types/xxi.htm.

This was the boat that perhaps could have won the war in the Atlantic for the Germans had she been in the water maybe 2 years earlier. She was the first real combat submarine that was meant to dwell in the deep and not just retreat to it once in danger.

3-times the electrical power of the VIIC gave the boat enormous underwater range compared to the older types and this boat could submerge far beyond the Bay of Biscay from the French bases so the Valley of Death was a thing of the past for them really. It took the boat 3-5 hours to re-charge the batteries with the Schnorchel once every 2-3 days if travelling at moderate 4-8 knots and was thus much less in danger from aircraft which sank about 56% of all U-boats lost in the war.

U-2540, now called Wilhelm Bauer, as seen in Bremerhaven.

As well as the advent of sonobuoys, airborne and seaborne radar, and ASDIC the other technologies and changes in tactics which contributed to countering U boats were:

• The closing of the mid-Atlantic gap with longer range MPAs such that the whole of Allied convoys transit routes had MPA cover. The result was that pre-snorkel early U boats were forced to spend more time submerged where they had limited mobility. The advent of the snorkel changed this, but air cover still had a deterrent effect.

• Professors Blackett's team's analysis of convoy losses over the first three years of the war showed that the overall size of a convoy was less important than the size of its escorting force. This was because large convoys were only slightly more detectable by U boats than small ones. It was therefore, more effective to concentrate escorts to better protect a few large convoys than many small ones.

- Escort carriers to provide the convoy with air cover, as well as close the mid-Atlantic gap.

- Bletchley Parks Enigma which enabled U boat tasking messages to be decoded such that their movements could be predicted

- Professor E J Williams' (ex-RAE, who was part of the OR team), analysis of why so many depth charge attacks did not damage U boats. He analysed U boat escape manoeuvres and concluded that the detonation depth setting should be changed from 120 ft (36m) to 20 ft (6m). Detonation depth being changed resulting in a tenfold increase in U boat sinkings.

- Professor E J Williams' recommendation that the underside of MPAs be painted a cloud colour to reduce the range at which they could be detected visually by U-boat crews.

- Prof Blackett's team member Cecil Gordon's maintenance and flying hours' study which tripled MPA effective flying hours.

- High frequency direction finding (HF/DF), including shipborne sets, to pinpoint the location of an enemy submarine from its radio transmissions.

- Professor E J Williams' study which led to MPAs deploying Leigh lights enabling radar detected surfaced U boats to be illuminated and surprised and attacked at night. This was needed as the MPA radar had a minimum range such that MPAs lost radar contact as they closed on a target at night.

- Air raids on the German U-boat pens at Brest and La Rochelle.

'SEA EAGLE' by J.S. Bailie
The first attack with the aid of a Leigh Light took place in the early morning of 4 June 1942 when Wellington VIII serial ES986 of 172 Squadron, flown by Sqn Ldr Jeaffreson H. Greswell from Chivenor in North Devon, was hunting off the north coast of Spain. The Italian submarine *Luigi Torelli* from Bordeaux, commanded by *Tenente di Vascello* Augusto Migliorini, was suddenly lit up and then badly damaged by four depth-charges. The submarine ran aground near Cape Peñas in Spain but was eventually freed. It was damaged further on 7 June by Sunderlands in a daylight attack. After beaching at Santander in Spain, it underwent temporary repairs and returned on the surface to Bordeaux on 14 June, escorted by relays of German aircraft.

Sunderland Strike painting by Mark Poselthwaite, eminent aviation
*(artist - www.posart.com)*

The table below summarises the progress of ASW during World War 2 and is reproduced from Willem Hackmann's Seek and Strike-Sonar, Anti-submarine Warfare and the Royal Navy 1914-54 published 1984.

It can be seen that when sonobuoys were introduced that the Battle of the Atlantic began to swing in favour of the allies, although there were also other factors which contributed including the introduction of the VLR (Very Long Range) version of the Liberator to close the mid-Atlantic gap.

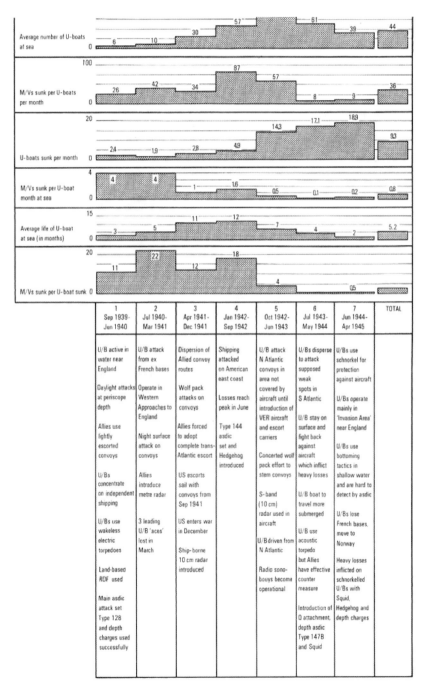

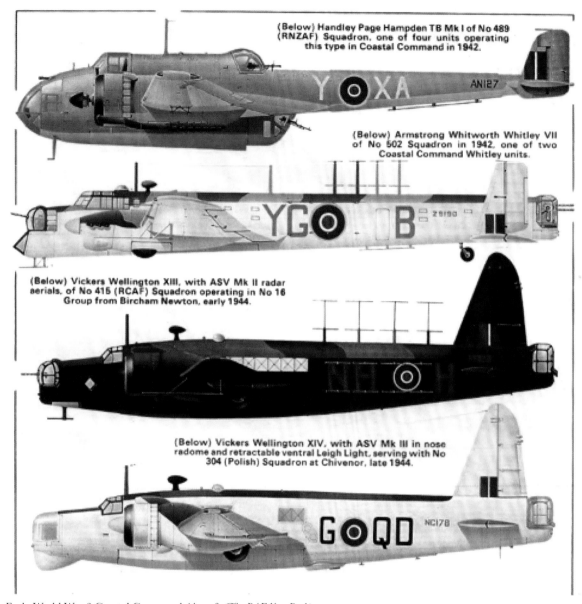

(Below) Handley Page Hampden TB Mk I of No 489 (RNZAF) Squadron, one of four units operating this type in Coastal Command in 1942.

(Below) Armstrong Whitworth Whitley VII of No 502 Squadron in 1942, one of two Coastal Command Whitley units.

(Below) Vickers Wellington XIII, with ASV Mk II radar aerials, of No 415 (RCAF) Squadron operating in No 16 Group from Bircham Newton, early 1944.

(Below) Vickers Wellington XIV, with ASV Mk III in nose radome and retractable ventral Leigh Light, serving with No 304 (Polish) Squadron at Chivenor, late 1944.

Early World War 2 Coastal Command Aircraft *(The RAF Year Book)*

The next chapter contains scans of letters/papers written in 1941 describing the initial concept of a sonobuoy as proposed by Professor P M S Blackett, UK scientist, and the subsequent correspondence with the USA re the concept.

# CHAPTER 3

## LETTERS/PAPERS FROM THE 1940S RELATING TO PROFESSOR BLACKETT'S IDEA FOR A DETECTOR BUOY AS HE FIRST DESCRIBED SONOBUOYS

What follows are scans of letters/papers obtained from the National Archives and written in 1941 describing the initial concept of a sonobuoy as proposed by Professor P M S Blackett, UK scientist, and the subsequent correspondence with the USA re the concept.

Professor Blackett was one of the pioneers of Defence Operational Research (OR).

Article about Patrick Blackett/*Intl. Trans. in Op. Res. 10 (2003) 405–407*

> Patrick Blackett is a towering figure in the history of operational research. He is, quite rightly regarded as the 'father of operational research' as a result of his membership of the Tizard Committee on Air Defence and his pivotal role in the creation of operational research sections attached to the principal British military commands after 1940. In that setting, his contribution to enhanced military effectiveness was profound, especially in relation to the defeat of the German U-boat campaign in the North Atlantic. His seminal papers emanating from his work on behalf of RAF Coastal Command and the Admiralty were characterized by a remarkable clarity of thought, reinforced by absolute intellectual integrity. This was especially the case in relation to his views on the appropriate location of operational research in military command structures and the emergent methodology of the new discipline. Blackett's powerful advocacy and effectiveness were noted in the USA and proved influential in diffusing operational research throughout the American armed forces after Pearl Harbour.

## Notes on proposed Detector Buoy
### by Professor P.M.S.Blackett.

The detecting device can be either a directional microphone or an ASDIC. The former gives bearings only, but is simple technically, and this should be the first to be developed. The latter gives bearing and range, but is markedly more complicated and had better wait for the moment.

One particular solution of the problem would be rather as follows :-

(i) The buoy should be stowable in the ordinary bomb racks of an aircraft and should be made as light and small as possible with, as an upper limit, the size and weight of a 500 lb. A/S bomb.

(ii) The buoy would be released from a low height e.g. 50 to 200 ft at speeds up to 170 knots. A parachute will be required to reduce the speed of entry into the water.

(iii) The buoy when floating on the surface, is rotated slowly and continuously in the water by, say, a small screw or paddle. The sound picked up by the directional microphone fitted to it, is amplified and transmitted by R/T from a low power transmitter and ariel, to the aircraft. The range of reception need only be 5 to 10 miles.

(iv) The buoy would be fitted with a magnetic compass which allows an R/T signal to be given automatically when the microphone reception direction is pointing North, say. Then the direction of the source of sound relative to the buoy will be determined by the time between the 'North' signal and the maximum sound from the source e.g., if the period of rotation of the buoy is one minute and the U/B is heard 17 seconds after a 'North' signal, the U/B must be on the bearing 108°.

Alternatively the instantaneous bearing might be transmitted by a frequency modulation device such as is familiar in meteorological sounding balloons.

The suggestions given above are only meant to give some concrete idea of what I have in mind. There probably are many variants which should be explored.

A sophistication - to be definitely left till the simpler problem is solved - would be to control the orientation of the buoy by W/T, so as to be able to keep it on the target.

I feel that it should not take very long for this buoy to be developed in the U.S.A. The Bell Labs. would find it easy I should think. The technique of the transmission of various magnitudes by R/T from sounding balloons has been highly developed in the U.S.A., by Johnson of the Bartol and many others.

18.5.41.

MINISTRY OF AIRCRAFT PRODUCTION,
THAMES HOUSE,
MILLBANK, S.W.1. 50 -33

anklin 2211.

Extn. 2260,

CENTRAL SCIENTIFIC OFFICE
JUN 1 1 1941
B. P. C.

24th May, 1941.

Dear Darwin,

    A problem has been put to me which might perhaps interest the N.D.R.C. and although I have not received any official request for work to be initiated I take this early opportunity of letting you hear of it.  It is probable it will come to me later as being a definite operational requirement but even if it does I doubt if we could tackle it effectively in this country with our resources so fully employed.

    I suggest, therefore, it might at once be put to the N.D.R.C. for consideration, with an enquiry whether they would care to put it on their programme.

    The idea is to provide a detector buoy to enable an aircraft to keep in touch with a submerged submarine. It is now fairly certain that the magnetic detector (known as M.D.S.) which Williams has been developing will not attain a range of much more than 200 ft. on a submarine.  The tactical value of the instrument is, therefore, definitely limited.

    There is, then, an urgent need to develop other methods by which an aircraft can keep in touch with a submerged submarine after it has been located.  The easiest way of attaining this is to drop from the aircraft a buoy which is fitted with an acoustic detecting device and which transmits the indication of the device by R/T to the aircraft.

    As you will have guessed the idea of the buoy emanates from Blackett and I send you herewith some notes giving his ideas in rather more detail.

/I....

Dr. C. G. Darwin, F.R.S.,
  Director of Central Scientific Office,
    British Supply Council in North America,
      725, 15th Street, N.W.,
        Washington, D.C.

19th June, 1941

Dr D. Pye,
Director of Scientific Research,
Ministry of Aircraft Production,
Thames House,
Millbank, London, S.W.1..

Dear Pye,

On the receipt of your letter of the 24th May about Blackett's scheme against submarines, I got in touch with the chief men of the N.D.R.C. and found they were extremely willing to take it up. As it happens, they had already got a project ~~in connection with~~ the *without* use of an airplane which embodied some of the same problems, and though they have hardly begun working on it they have given considerable thought to this other problem. Much of this can be taken over to the much more difficult problem of Blackett's, and they have already begun to think about it.

It is evidently not going to be an easy business, but I am sure that they are going to try very hard, and I will keep an eye on what they do and let you know in due course.

It looks as though the central difficulty would be the direction finding, since the hydrophone in the acoustic range is obviously the easiest and most suitable, but this has very poor directional qualities. I have told them that it seems to me that if they could even get the direction within 45° it would be a great deal better than nothing. I will let you know how the matter progresses.

Yours,

C. G. Darwin,
Director,
Central Scientific Office.

CGD:MEM

The US are keen to take up the idea

50 - 3

### NATIONAL DEFENSE RESEARCH COMMITTEE
#### OF THE COUNCIL OF NATIONAL DEFENSE
#### 1530 P STREET NW.
##### WASHINGTON, D. C.

NEVAR BUSH, CHAIRMAN
HARD C. TOLMAN, VICE CHAIRMAN
R ADMIRAL H. G. BOWEN
WAY P. COE
S. T. COMPTON
ES D. CONANT
NK B. JEWETT
. GENERAL R. C. MOORE

N STEWART, SECRETARY

11th Floor
172 Fulton Street
New York, N. Y.

June 30, 1941.

Dr. C. G. Darwin, Director
British Central Scientific Office
725 - 15th Street, N.W.
Washington, D. C.

Dear Dr. Darwin:

   Thank you for sending the blueprint of the
500 lb. bomb. For the moment, I think we may assume
that the buoy can be so designed as not to exceed the
sizes specified. Indeed, I hope that it will be possible
to do much better than that by providing some kind of
extensible envelope which will automatically inflate
when the buoy is dropped. The usefulness of the device
will be so tremendously enhanced by making it small and
light that we shall make every effort to reduce these
factors to a minimum.

      Sincerely yours,

      John T. Tate,
      Vice Chairman, Division C.

Copy to Dr. Shea.

CENTRAL SCIENTIFIC OFFICE
JUL 2 1941
B. P. C.

Please quote: 50-3-5

27th October, 1941

Dr C.S.Wright,
D.S.R.,
Admiralty,
Dorland House,
Lower Regent St.,
London, S.W.1.


Dear Wright,

I have answered your cable about the sonic buoy to the best of my ability. You understand that this buoy is the first idea of Tate's as a general watchman behind a convoy, and they have not yet tackled the much more difficult question of making the microphone directional. I am hoping to see them during the coming week, and will probably be writing you a report on any ideas they have on this more difficult problem.

In the meanwhile, however, I enclose a report which will give your people plenty to think about. I have only one comment to add, and that is that you may not realise that what they call rough water here would not be called rough water in the seas round England, at least I judge this from the rigging on their yachts and that they seem to regard what we would call a stiff breeze as almost a gale.

Yours sincerely,

C.G.Darwin,
Director,
British Central Scientific Office.

Enc:"Progress on the development of an expendable radio sonic buoy."
"Report on NDRC Contract No. OEM-sr-33."

CGD:MEM

T-10197,


Letter re Progress

## PROGRESS ON THE DEVELOPMENT OF AN EXPENDABLE RADIO SONIC BUOY

A contract was drawn up with the RCA Manufacturing Company to design and construct two models of an expendable buoy which would support an underwater microphone which upon receiving sound would modulate a small radio transmitter in the buoy.

It was suggested originally that such a buoy might be useful in convoy work. Buoys could be dropped at intervals behind the convoy and would indicate by radio to the convoy the presence of a submarine approaching from the rear. It was hoped that if the buoy could be made small and light enough that it might also be carried by airplanes and dropped in the neighborhood of a submarine. It would then relay to the airplane by radio any sounds picked up from the submarine.

A fundamental factor in the design of the large type buoy to be used from ships is its cost. By making the buoy in quantities it is hoped that this cost can be kept down to about $100 each. The buoy is to contain a soluble plug so that it will sink after a few hours' use.

Two experimental buoys have been constructed and working drawings prepared. Preliminary tests have been made at the New London laboratory. In these tests sounds from surface ships were picked up at a distance of several miles. In shallow water difficulties occurred because of the masking effect of noise produced by movement of the water on the bottom. In deeper water it has been possible to pick up a ship which was about eight miles distant.

No measurements of temperature gradient were made in these tests so that the range of the supersonic equipment under the same conditions is not known. The radio range of the buoy has been found to be approximately twenty miles although those working with it think the range can probable be extended somewhat further.

In the preliminary tests although high waves broke over the buoy no appreciable fading of the signal occurred. Additional tests of the buoy are to be made during the next two or three weeks.

**CONFIDENTIAL**

Progress Statement

LORSA 218


To:     Admiralty

From:   British Purchasing Commission

For D.S.R. from Darwin


Reference SALOR 1954

Sonic buoy has been tried at sea.  It gives no indication of direction of sound, but is intended to serve as watchman behind convoys.  Radio range 20 miles and sound range estimated 8 miles for deep water.

Microphone 30 ft. down.

Buoy about 2 ft. long and 1 ft. diameter and sinks after 3 hours.

Antenna 3 ft. long, radio frequencies round 70 megacycles in 3 distinguishable types.  It has continued working in fairly rough sea.

Estimated cost $100.

Full report being forwarded.


Central Scientific Office          2 (Originator:  Dr. Darwin)

Trial Outcome 1941

Chapter 4 describes how Professor Blackett's sonobuoy idea was developed into an operational system.

# CHAPTER 4

## THE FIRST OPERATIONAL SONOBUOY, THE AN/CRT-1, USED BY THE USA AND THE UK IN WORLD WAR 2

This chapter was compiled from a number of sources including Roger a Holler's T*he Evolution of the Sonobuoy from World WAR II to the Cold War* published in 2014, U.S. Navy Journal of Underwater Acoustics, Jan 2014. He in turn partly drew on Russell I Mason's paper, Sonobuoys Historical Development, published in an IEEE Newsletter in 1984 and other sources.

The joint war effort between the UK and USA served to facilitate early sonobuoy development activities with concurrent investigations occurring at the War Research Laboratory of Columbia University, the Sonar Research Unit of Portland, and the Royal Aircraft Establishment at Farnborough. This joint effort was initially aimed at producing a surface ship launched sonobuoy. The concept of air deployed sonobuoys had been British but they could not spare the technical resources to develop the idea at that time and the US led the initial development effort. RCA Camden and Columbia University at New London, Connecticut carried out initial development work in a joint war effort with the Sonar Research Unit Portland and the Royal Aircraft Establishment Farnborough.

RCA delivered the first models of a proof of concept ship-deployable buoy in less than three months. The small, 60 lb sonobuoys were tested on September 12, 1941, at Barnegat Bay in New Jersey and proved the concept of an expendable sonobuoy. Despite typical teething problems such as they tended to leak, sank and were noisy they proved the concept

The US Navy then cancelled the programme to pursue other, higher priority efforts, but in early 1942 the air dropped sonobuoy concept was taken up. A rapid redesign took place and then the Army Air Corps expressed interest in aircraft launch. Airborne tests using Blimp K-5 and the US Navy Submarine S-20 validated the overall concept of employing sonobuoys for detecting submarines. On March 7, 1942, two RCA ship-launched sonobuoys were monitored from a blimp as they tracked the S-20 submarine off New London, Connecticut.

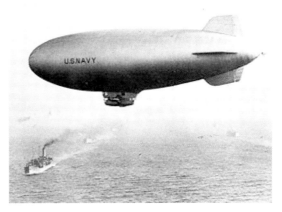

Blimp K-5

Progress was rapid and the first aircraft air drop was from an US Army B 18 bomber in July 1942. One month later, buoys were dropped against a U boat and shortly afterwards the US, UK and the Canadians were using them to detect U boats.

While the first airborne sonobuoys were hand-made by engineers and technicians in New London, larger quantities required commercial manufacture. The first contract to General Electric, Bridgeport, Connecticut, for 22 buoys was a failure. Subsequently, Freed Radio and Emerson Radio, both of New York City, produced sonobuoys in large quantities, based on previous designs and working closely with Underwater Sound Lab engineers. Freed also manufactured the FM receivers.

Once dropped from an aircraft, a parachute deploys to slow the descent of the sonobuoy. On impact with the sea it separates into two parts connected by a cable. One part containing a hydrophone and associated electronics sinks to a predetermined depth set by the length of the cable. The other is buoyant, remains on the surface and carries a radio transmitter and antenna to provide a link to the aircraft. Buoyancy in the early sonobuoys was provided by a rigid cylinder. This system was replaced in the late 1970s by small gas inflated flotation bags as a result of a drive to reduce the size of sonobuoys.

The Douglas B 18 was the first aircraft to drop a sonobuoy

The first air-droppable sonobuoys were packaged in a three-foot-long cylinder about five inches in diameter, a size that later became standardised as A-size. Avoiding the use of materials critical in wartime, the outer cylinder for these first buoys was a quarter-inch-thick paper tube coated to keep it watertight for a few hours. Wooden disks at the ends of the combined electronics and battery compartment were sealed by adhesive tape and flexible pitch, providing watertight integrity. It had a half-watt radio transmitter tuned to one of six FM frequencies, used five vacuum tubes and had a steel monopole antenna for transmitting the underwater sounds received by the hydrophone to a receiver (AN/ARR-3) in the aircraft.

The radio range was approximately 10 miles when the aircraft was at 300 feet. The underwater range was in the region of 200 to 3,000 yards depending on the water conditions. Power was provided by flashlight batteries. Its 24-foot-depth non-directional hydrophone was a broadband listening device made of rugged nickel magneto-striction material.

The resulting sonobuoy was the AN/CRT-1, a passive omnidirectional broadband detection device. The AN/CRT-1 was 40 inches long and weighed 14 lb.

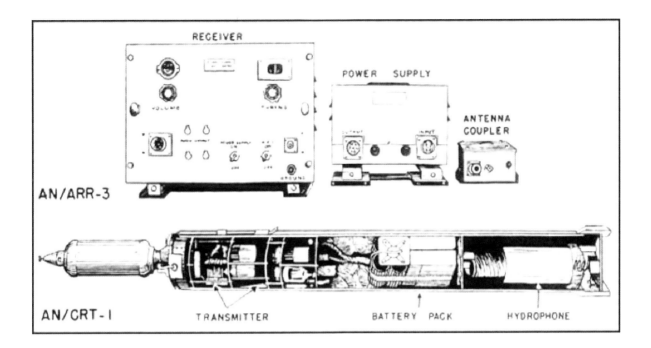

In August 1942, the AN/CRT-1 became the first operational sonobuoy. By the end of October 1942, the procurement of the expendable radio sonobuoy for use in anti-submarine warfare had been initiated. The Commander in Chief, U.S. Fleet, directed the Bureau of Ships to procure 1,000 sonobuoys and 100 associated receivers. The Army Air Corps ordered 6410 of the AN/CRT-1 passive sonobuoys, while the Navy ordered only 1800 units. In 1943, full-scale sonobuoy production was started in the United States by RCA under the direction of the Office of Scientific Research and Development.

The USAAF became enthusiastic about the sonobuoy and ordered over 6,400 of them in 1942. The AN/CRT-1 sonobuoy was in formal service by 1943, initially carried by USAAF Liberator ocean patrol aircraft out of Newfoundland and the UK. The US Navy hadn't followed through on the blimp experiments, but the USAAF experience convinced the Navy that the sonobuoy was a good idea, and the Navy bought almost 60,000 AN/CRT-1As during the war.

The first use of sonobuoys against a U boat resulting in a sinking was by a Canadian RCAF Hudson operating out of Iceland in 1943. The U boat was caught on the surface and a pattern of sonobuoys were dropped round it. A depth charge attack was successful and recordings from the sonobuoys captured the sounds of the submarine breaking up.

AN/CRT-1 sonobuoys produced in the USA during World War 2 were used in action around the world and were rather basic by modern standards but played an essential role in the birth of Airborne ASW. Early in 1943 the US Navy deployed the system in B-24 Liberators flying from Argentina, Newfoundland, and from Dunkeswell, Devon. By the end of the war over 150,000 sonobuoys had been produced.

AN/ARR-3 Aircraft Receiver

The AN/CRT-1 could transmit on one of six radio channels. Each channel corresponded to a colour (purple, orange, blue, red, yellow, and green) and was received on the AN/ARR-3 receiver, which the operator manually tuned to one buoy at a time using a colour-coded tuning window to compare the intensity of the sound. Automatic Frequency Control (AFC) allowed rapid tuning for comparative listening.

A typical attack procedure would be that the aircraft would drop a purple sonobuoy on the suspected position of a submarine and follow with four other colour buoys, deployed as the aircraft flew in a cloverleaf pattern by over-flying the purple buoy on each leg of its run, the aircraft could position the other buoys, one per pass, at a distance of 2 miles from it. The usual order was Purple, Orange, Blue, Red, and Yellow (POBRY), with green being reserved as a backup. When an aircraft had a contact on a surfaced submarine, it would approach at 300 ft, dropping a weapon and one or more buoys to listen for the explosion or the cavitation of the submerged submarine's propellers. If the kill was not confirmed, additional buoys would be dropped to relocate and track the target.

The AN/CRT-1 was most effective in detecting submarines traveling at cavitating speed, i.e. the boats propellers were rotating at a sufficiently high speed to release bubbles of air, which popped to give a distinctive sound. The speed at which cavitation occurred varied with depth, the deeper the depth the faster the submarine could travel without cavitating. When a submarine ran

R-2/ARR-3 Aircraft Receiver for Sonobuoys and Color-coded Tuning Window

underwater slowly, and its propellers weren't cavitating detection ranges were very short, e.g. about 90 yards against a submarine at 3 knots at a depth of 250 ft in a rough sea. Whereas in calm sea conditions against a seven-knot submarine at 60 ft much higher detection ranges were possible.

As the AN/CRT-1 was a passive listening sonobuoy with an omnidirectional hydrophone, precise targeting was difficult. It was laid in patterns of five buoys and the aircraft operator estimated the signal strength of the target on each of the buoys in contact. The navigator then used concentric circle overlays in an attempt to localise the submarine using relative signal strength.

This early method of localisation proved to be rather inaccurate and led to the development of a directional sonobuoy, the AN/CRT-4, but it was too late to be used in the war.

A number of RAF aircraft used the AN/CRT-1 during World War 2 including the Sunderland and the US supplied Liberator.

Prototype UK Short Sunderland 1938

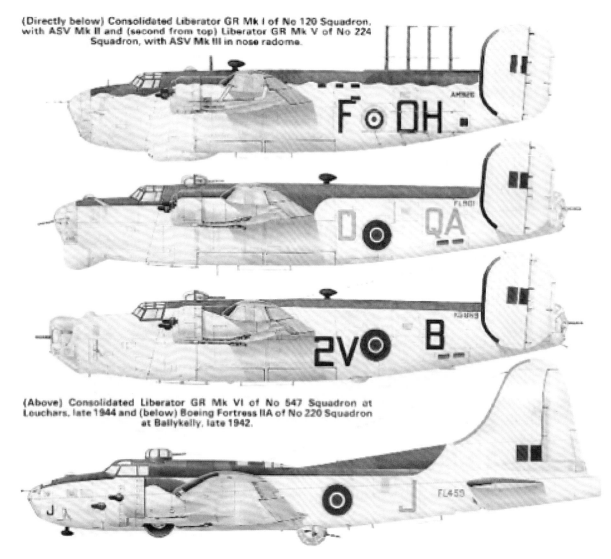

(Directly below) Consolidated Liberator GR Mk I of No 120 Squadron, with ASV Mk II and (second from top) Liberator GR Mk V of No 224 Squadron, with ASV Mk III in nose radome.

(Above) Consolidated Liberator GR Mk VI of No 547 Squadron at Leuchars, late 1944 and (below) Boeing Fortress IIA of No 220 Squadron at Ballykelly, late 1942.

VLR World War 2 Coastal Command Aircraft (The RAF Year Book)

The following paragraphs are a summary of a Growler magazine article

The year 1943 represented the turning point in the Battle of the Atlantic partly because of the closing of 'The Mid-Atlantic Gap'. Until the appearance of the Liberator a whole swathe of the Atlantic was effectively uncovered giving the U-boast free rein. By 1943 120 Squadron was operating the VLR Liberator V. VLR stood for Very Long Range which was achieved by additional fuel tanks in the bomb-bay and by the removal of much of the armour.

Seen from the Allied point of view, the Atlantic war was potentially fatal for Britain and until the arrival of Coastal Command Squadrons with VLR aircraft the U-boats were winning.

After the war finished the UK downsized its airborne ASW assets and introduced some new aircraft. The aircraft changes are described in the next chapter.

# CHAPTER 5
## THE UK UPDATE THEIR FIXED WING ASW MPAS IN THE 1950S

Most of the text relating to the Shackleton was compiled from the Shackleton Association web site.

The majority of Liberators in RAF service were obtained under the US/UK Lease/Lend deal so with the end of the war, nominally, they had to be handed back. In reality most of them were simply melted down.

Lancaster GR 3.

The Liberator was replaced by the Lancaster GR 3. Although the Lancaster was a good aircraft, the GR3's had simply been modified from World War bombers and were extremely limited in the available space not just for larger maritime crews but also for the large amount of equipment demanded for maritime operations. But even the Liberator with its much larger fuselage had originally been designed as a long-range bomber. What Coastal Command needed was an aircraft with the space of a Liberator but specifically designed for maritime reconnaissance/bomber operations. What resulted was effectively, the Lancaster successor's (the Lincoln) wings married to a Liberator fuselage, the Avro Shackleton.

To help fill the gap when the Liberators were scrapped the UK purchased some US Neptunes as an interim measure.

Neptune - *Nimrods Genesis*

Unlike the UK's other MPAs the Neptune did not have the underside of its fuselage painted in a light colour to minimise visual detections by the crews of surfaced submarines.

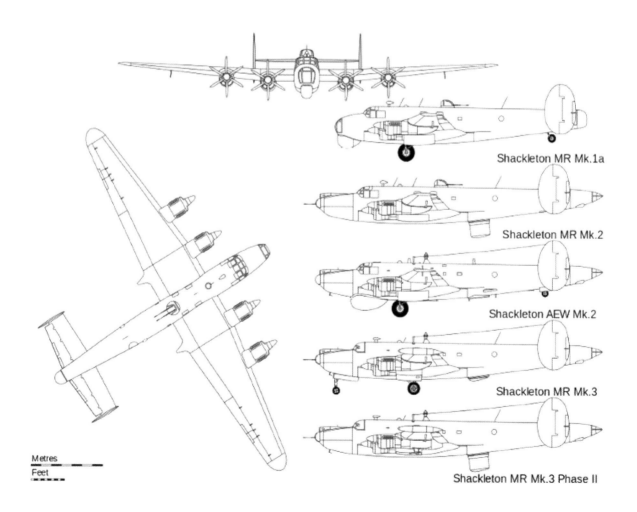

Shackleton MR Mk.1a

Shackleton MR Mk.2

Shackleton AEW Mk.2

Shackleton MR Mk.3

Shackleton MR Mk.3 Phase II

Metres

Feet

Shackleton MR1 *(Stock picture)*

Originally to be known as the Lincoln Mk 3, due to its main Spar and wing ancestry, but fatter and shorter and much more powerful, it was named the Shackleton by Roy Chadwick, who had joined the infant A V Roe as a draughtsman in 1911 aged 18, and rose to become Chief Designer, collecting the CBE, and qualifying as an MSc, ERSA, FRAeS and AMCT over the years. The first prototype VW126 made its first flight on 28 March 1950.

Prior to the decision to adopt the Shackleton, the UK flying boat manufacturers had a mounted a vigorous campaign for a new design flying boat to be UK's next Maritime Patrol Aircraft. However, after numerous studies the UK decided that no more flying boats were to be developed for military purposes. Flying boats tended to be less efficient as aircraft and needed numerous boats etc. to support them and were only considered by the MOD because of the UK's east of Suez commitments where it was envisaged a Flying boat would be required.

Seventy-seven MR1 and MR1A aircraft were built, including the prototypes, with production ending in July 1952. A FAST volunteer flew in Shackletons in the mid-1950s. He remembers a mission in the NE Atlantic searching for Soviet Union submarines. They found a conventional SSK on the surface and it dived immediately. They then deployed 2 passive sonobuoys and a few minutes later the SSK surfaced again right by the sonobuoys. A submarine crew member then appeared and pulled one of the sonobuoys on board the submarine. He had a quick look at it and then threw it back into the ocean. It appeared the Soviets must have already had a number of sonobuoys of that type and didn't need another one.

A total of 185 Avro Shackleton aircraft were built between 1951 and 1958 and it continued in the Maritime Reconnaissance and Search and Rescue roles until 1972. Twelve MR Mk2 aircraft were converted to AEW.2 and continued in operational service until June 1991. The RAF were never completely happy with the Shackleton as it never fully met their requirements.

Most Shack men accepted that it was not really an ideal aircraft from the crew point of view for 15 hours plus flogs. Noisy, gloomy, ergonomic nightmares, difficult to land and, the source of 'informed criticism', physiological tests showed that crews were comfortable on less than 1% of their entire trips. Investigations by the Institute of Aviation Medicine found that intensive flying in the Mark 2s caused great loss of efficiency, buzzing in the ears and sleeplessness (when not flying, one presumes).

The following is a part of an article from Royal Airforce Historical Society Journal 33 written by Air Cdre Bill Tyack retired. The article describes a typical Shackleton ASW exercise. He is at the bottom of the picture below during a Sunderland, mission

Sunderland navigator Bill Tyack bottom of Picture-*Flypast magazine*

# ANTI-SUBMARINE ACTION

Radar was our main anti-submarine search sensor. The task was to search a patrol area with radar s:) as to detect a submarine during its relatively short snorkelling period, typically 20 minutes every few hours in the more modern boats A submarine snorkel would offer a target echoing area of about one square metre which ASV 21 might detect at up to 15 Nm (nautical mile) in very favourable conditions but at much shorter range in the sea states normally experienced in the North Atlantic. Another difficulty was that submarines hard radar intercept equipment that could detect ASV radar at a much greater range than the radar could detect the submarine. So, it becarne a game of cat and mouse, on top of hunting for a needle in a haystack. We used various tactics such as switching the radar on intermittently for short periods and/or scanning it in a sector behind the beam of the aircraft, to counter the range advantage enjoyed by the submarine. The hope was that we would detect the submarine before it had time to submerge in reaction to intercepting the aircraft radar. We could then home onto the radar contact for a direct attack or, if it had submerged, lay sonobuoys on the datum to relocate and attack. we practised this endlessly, homing onto radar buoys located new the man maritime bases or onto a skid target towed behind RAF marine craft that produced a wake similar to that of a snorting submarine.  Each homing culminated in a visual attack with practice bombs, aimed by the pilot from a height of 100 feet in daylight or the bomb-aimer from 300 feet at night.

Let me take you through a typical action. It is a dark night over the North Atlantic. A Shackleton crew has been airborne for eight hours on a radar search for conventional submarines that intelligence suggests are transiting through the aircraft's patrol area. it has been a quiet, boring petrol, broken only by endless cups of coffee, a fry-up shortly after take-off and a nourishing helping of 'Honkers Stew' a couple of hours ago. The crew is starting to think about the long transit back to base. Suddenly the radar operator reports a small radar contact that he assesses as a possible submarine. The captain calls 'Action stations, action stations turning on and the crew is galvanised into action. The first pilot turns onto the contact and homes on under the direction of the radar operator. The co-pilot selects maximum boost on the Griffons and the flight engineer starts the Viper engines to give additional power for safe manoeuvres at low level. The W/T operator sends a POSSUB message with details of the contact. Meanwhile the tactical navigator sets up his plotting table to follow the homing and take over control to the datum if the contact disappears. (If the contact disappears this offers some collateral evidence that it might be a submarine.) He checks the settings on the weapon control paid, sets up the sonobuoy pattern on his plotting table and confirms the sonobuoy serial numbers (equating to radio frequencies) with the sonics operators the routine navigator goes to the nose and checks the low-level bombsight.

The lookout positions in the tail and in the beam will be manned. At three miles the pilot selects the bomb-doors and camera doors open. The radar scanner is lowered to the attack position. The pilots gradually descend to the attack height, paying close attention to the radar altimeter. At one mile a sequence of flares is fired to illuminate the target. These are 1.75- inch calibre

Two Mk 101 Lulu nuclear depth bombs either side of a UK Mk 30 torpedo in the bomb bay of a Shackleton Mk1
*Photo Tim McLelland Collection*

pyrotechnics fired from dischargers in the beam upwards and to the side of the aircraft. If the bomb-aimer sights the target he gives directions to the pilot and releases the weapons in the early years this would have been a stick of depth charges, but in the 1960s it would be a passive Mk 30 homing torpedo and a Mk 44 active horning torpedo. Releasing the weapons also fires a series of six 1.75-inch photoflashes and triggers the K24 camera to record the results of the attack. The observer in the tail reports the results of the attack. An active Mk 1C sonobuoy and a smoke flame maker are dropped with the weapons to enable relocation and re-attack of the target, if necessary. If the bomb aimer does not sight a target, only the sonobuoy and marker are dropped on the datum, followed by a passive sonobuoy 2,000 yards further on. The pilot then flies the aircr4t in a teardrop, overflies the datum sonobuoy using the sonobuoy homer, which feeds signals to the zero-reader display. On top of the datum sonobuoy the navigation plot is updated and 2000 yards beyond another passive sonobuoy is laid. The crew then track the target using bearings on the submarines radiated noise from the passive sonobuoys and fixes from the active sonobuoy. Other sonobuoys are laid as required. Once the tactical navigator has an attack solution he steers the pilot to a drop point and releases homing torpedoes ahead of the target.

This scenario is entirely imaginary and was played out in countless times in exercises and training sorties.

During the Cold War nuclear depth bombs might have been deployed in Shackletons if the Cold War had turned hot.

(Below) Avro Shackleton GR Mk 1 of No 120 Squadron at Aldergrove, after adoption of the overall grey finish in succession to the predominantly white finish.

(Below) Hawker Siddeley Shackleton MR Mk 3 of No 201 Squadron, with which this variant operated at St Mawgan from October 1958 until October 1970.

Shackletons – *RAF Year Book*

# CHAPTER 6
## POST - WORLD WAR 2 UK SONOBUOY RESEARCH AND DEVELOPMENT

## BACKGROUND

After World War 2 UK design expertise and sonobuoy manufacturing facilities came into being both in MOD and the UK defence industry. The UK went on to be a major sonobuoy manufacturer along with the USA. The development of UK expertise was a reaction to the emergence of the Soviet Union as a potential opponent after World War 2 and the development of a significant submarine arm in its Navy. The following paragraphs in italics give the background to this development and are largely based on a U.S. Navy Journal of Underwater Acoustics article published in January 2014 - *The Evolution of the Sonobuoy from World War Two to the Cold War* by Roger A. Holler Navmar Applied Sciences Corporation

After World War II, the Allies sank or destroyed most of the remaining German U-boats, but some were divided among the victors. The Soviet Union received 10 of the German U-boats, including 4 of the most advanced Type XXI submarines. By January 1948, the Soviets had produced their own submarine and had 15 Type XXIs in their fleet. From June 1948 to May 1949, the Soviet Union's blockade of Berlin triggered the Berlin airlift and inflamed Western fears of the USSR. The outbreak of the Korean War in 1950 marked the true beginning of the NATO -Soviet Cold War. The Soviet submarine fleet grew in the 1950s with Zulu and Whiskey type submarines that were based largely on the German type XXI U-boats. In 1962, the Soviets produced the Juliett class of diesel electric SSGs (Guided Missile Submarines) with anechoic rubber tiles on the hull and announced that their nuclear submarines had fired ballistic missiles from submerged positions.

The Soviet Union deployed its first Hotel-class SSBN on a 70-day patrol into the Atlantic in April 1962. By October 1962, the Cold War reached crisis stage when surveillance flights showed evidence of nuclear missile installations and Soviet missiles on Cuba, leading to the Cuban missile crisis. During the crisis the Soviets deployed four Foxtrot diesel-electric attack submarines from the Kola Peninsula to the Caribbean Sea armed with torpedoes with nuclear warheads All 4 Foxtrots were eventually detected while snorkelling (3 by aircraft) and then were tracked and harassed and were forced to surface for battery recharging. All of them returned to base without making it to Cuba.

# THE IMMEDIATE POST WORLD WAR 2 PERIOD

The UK's early indigenous sonobuoys had a T number designation system. In reading the original papers at the National Archives I found the way the numbers were allocated then at times very confusing so forgive me if this section is also a little confusing.

| UK Sonobuoy Type | UK Sonobuoy Designation | Radio Frequency | Acoustic Frequency c/s | Dimensions inches | Wt. lb |
|---|---|---|---|---|---|
| Passive Omni | T-1945 | 62.9/71.7 FM | 200/5000 | 34" x 5 7/8 | 21 |
| Passive Directional | T-1939 | 60.15/62.35 AM | 12000 | 54" x 8" | 70 |
| Passive Directional | T-1946 | 60.15/62.35 AM | 12000 | 54" x 8" | 70 |

Summary of the first UK produced sonobuoys

At the end of World War 2, the only sonobuoy in use by UK, US and Canada was the US AN-CRT -1A. By the late 1940s the UK had virtually used all its stock of AN-CRT-1A sonobuoys as a result of training and other exercises carried out by the then UK training centre at the Joint Anti-Submarine School at Londonderry. To avoid UK stocks running out a UK non-directional sonobuoy, the T1945 was designed by the RAE at Farnborough and was comparable to the AN-CRT-1A in dimensions and weight and functionally interchangeable with it. Ultra-Electronics Ltd and McMichael Ltd started to produce these sonobuoys in numbers, in the later 1940s, early 1950s. It was a passive omnidirectional buoy operating in the 62 MHz to 72 MHz band and was supplied to both the Royal Navy and the Royal Air Force.

After the T1945, McMichael Ltd designed and produced 500 pre-production models of an alternative passive omni directional sonobuoy, designated the T1302, operating in a higher frequency band than the T1945. The first models were free-falling and, although this method of dropping was effective from moderate height, when dropped from low altitude the buoy tended to "porpoise" and either break up or malfunction. The solution was in parachute deployment, but the T1302 never went into quantity production.

In conjunction with the sonobuoy programme McMichael designed and produced a whole range of special-to-type test gear for sonobuoys and, as the equipment came into use in RAF Coastal Command aircraft, it built a series of synthetic trainers which were installed at various RAF stations throughout the world for the training of complete aircrews in the use of airborne submarine detection equipment.

McMichael Limited in 1961, became a largely autonomous member of the UK's General Electric Company (GEC) consortium of Companies and G.E.C. and gained its future head in Arnold Weinstock a famous UK industrialist.

T1945 - *EAST collection*

T1945 Early 1950s - *FAST collection*　　　　　　　　　　　　T1945 - *FAST collection*

Trials conducted by the UK after the war showed that localisation accuracy was very poor with the first generation omnidirectional sonobuoys. As a result, development of the first truly British sonobuoy began under the design and production responsibility of the RAE. Previously U.D.E had obtained twenty US directional sonobuoys the AN-CRT-4 and its receiving equipment the AN-ARR-16B. It was based on the AN-CTR-1A and had a directional hydrophone and a magnetic compass to enable the orientation of the hydrophone to be transmitted to the monitoring aircraft. UDE trials showed the AN-ARR-16B suffered from self-noise problems. In 1948 after evaluating the US directional sonobuoy the UK began development of a directional sonobuoy, the T1939, with RAE Farnborough having overall design and production responsibility. Hydrophone design was carried out by HUMDE and the Admiralty Compass Observatory developed the compass.

This passive directional sonobuoy, the T-1939, was 54 inches long and weighed 70 lb and early production models were available in 1949 for testing. Directionality was achieved by rotating the listening hydrophone at 3 rpm using an electric motor driven by kalium cells. The torque was taken by a system of three paddles. The hydrophone was designed at the HMUDE and consisted of an array of 12 wound nickel rings inserted in resonant cavities in a cruciform shaped plate. The hydrophone listened over a kHz band centred at 12 kHz at a depth of 40 ft.

The detection display was a CRT with a linear bearing scale arranged horizontally across the face of a CRT with a spot indicating the orientation of the hydrophone.

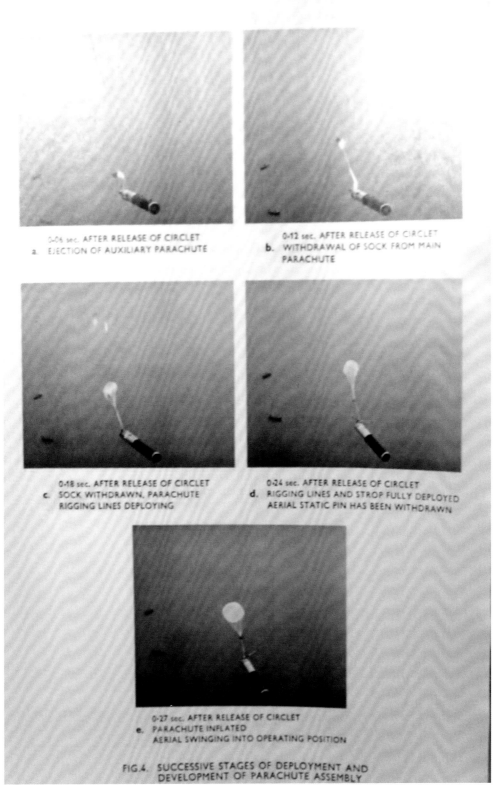

0-06 sec. AFTER RELEASE OF CIRCLET
a.  EJECTION OF AUXILIARY PARACHUTE

0-12 sec. AFTER RELEASE OF CIRCLET
b.  WITHDRAWAL OF SOCK FROM MAIN
PARACHUTE

0-18 sec. AFTER RELEASE OF CIRCLET
c.  SOCK WITHDRAWN, PARACHUTE
RIGGING LINES DEPLOYING

0-24 sec. AFTER RELEASE OF CIRCLET
d.  RIGGING LINES AND STROP FULLY DEPLOYED
AERIAL STATIC PIN HAS BEEN WITHDRAWN

0-27 sec. AFTER RELEASE OF CIRCLET
e.  PARACHUTE INFLATED
AERIAL SWINGING INTO OPERATING POSITION

FIG.4.  SUCCESSIVE STAGES OF DEPLOYMENT AND
DEVELOPMENT OF PARACHUTE ASSEMBLY

T1939 *National Archives*

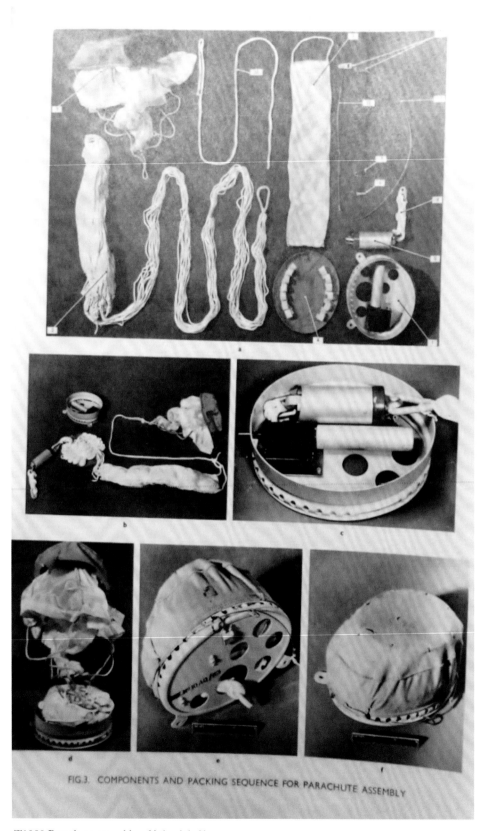

FIG.3. COMPONENTS AND PACKING SEQUENCE FOR PARACHUTE ASSEMBLY

T1939 Parachute assembly - *National Archives*

T-1946 SONOBUOY- picture dated October 1951- *EAST collection*

In the early 1950s the USA, UK and Canada entered into an agreement to standardise the main features of sonobuoys to ensure interoperability. This covered the physical characteristics i.e. size and weight and the RF link. Later other countries joined this agreement.

The standard sizes defined were:

A- 4 and 7/8 ins diameter, 36 inches' long
B- 6 inches' diameter, 60 inches' long
C- 91/2 inches' diameter, 60 inches' long

In 1954 the T-1946 was redesigned by Ultra to operate on the new tripartite (UK, USA, and Canada) specified 16 radio channels to enable interoperability of sonobuoys between NATO nations and was known as the T7725. The T7725 was 5 ft long 8 and 3/4 in diameter and weighed 70 lb and was bought by the US Navy as the AN/SSQ-20 as their own directional sonobuoy the AN/CRT-4 was unsuccessful. In the UK only a 1000 T7725s were produced.

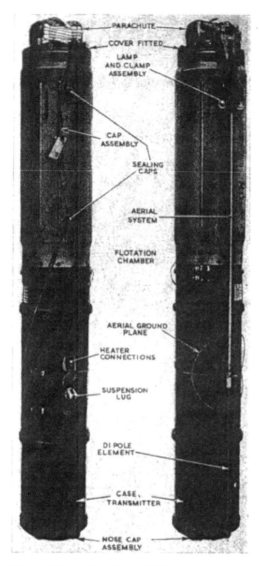

**Fig. 1. Transmitter Type T.7725—general views**

3. In its pre-drop state all of the units of the sonobuoy are contained in a metal canister (fig. 1) suitable for fitting on a light series bomb carrier. The construction of the sonobuoy permits it to be dropped from an aircraft flying at a speed of up to 150 knots at a minimum altitude of 350 ft. or at speeds of up to 250 knots at altitudes between 500 ft. and 10000 ft. When released from the aircraft, the rate of descent is controlled by a small parachute.

4. On impact with the sea, the transmitter sub-assembly Type M1 (known as the lower or submerged unit) is released and slides out of the canister, sinking to a depth of 32 ft. determined by the length of the interconnecting cable. The upper section of the cannister itself (the flotation chamber main assembly) is sealed and remains buoyant with a freeboard of about 4 in., suspending the submerged unit at the required depth.

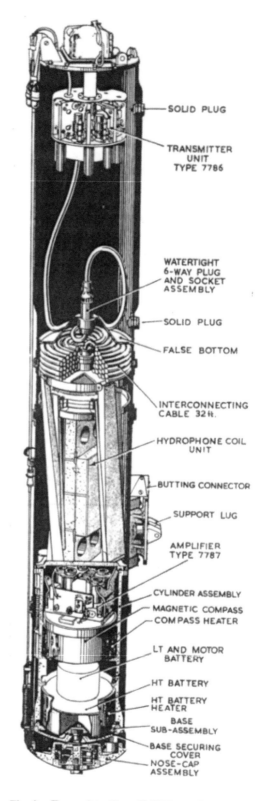

**Fig. 2. Transmitter Type T.7725—sectional view**

RAE Andover deploying sonobuoys at a test range  *Ultra Electronics picture*

The T7725s beam width for half sensitivity (6 dB down level) was 36 deg and it could operate for 1 hour using two batteries and float for 4 to 6 hours. In its steel cylindrical canister 2 ft 6 in from top was a false bottom which with the top plate formed a watertight compartment. It had two 2 soluble plugs which used sodium bicarbonate. The radio uplink had a range of 10nm (nautical mile) for an aircraft height of 500 ft. and operated between 162.25 and 173.5 Mc/s on 16 transmission channels.

A drive motor operated via reduction gearing against the reaction of the paddle unit to rotate the main body of the dumb merged unit at 3 rev per minute.

Early buoys needed a separate marine marker and smoke and flares to show buoy positions.

## THE FIRST UK ACTIVE (PINGING) SONOBUOY

In the 1950s MOD was planning a directional active sonobuoy system to enable an accurate submarine position to be established prior to weapon launch which was known as the MK 1 Sonics system. The sonobuoys used in the Mk1 Sonics System comprised a monostatic active buoy(T1154) and a passive buoy(T9003). The buoys were very large because they each included a mechanical assembly and motor to physically rotate their projector and/or hydrophone. The T9003 directional receiver rotated mechanically at 3 rpm.

The T1154 unlike the T9003 did not rotate continuously, but in series of ten steps with a pause of about 3 seconds to listen for return echoes. The T1154 required a transducer that could act both as a projector and receiver. The chosen system consisted of 4 short barium titanate cylinders assembled to form a hollow air-filled tube about 4.5 inches long, giving a half wavelength at the operating frequency of 20.4, 21.7 or 23 KHz. Vertical directivity was obtained by a foam rubber coated parabolic reflector. The acoustic pulse was generated by sub miniature valves which operated in a resonant push-pull circuit at 450 volts HT which was well in excess of their rating.

At this time all sonobuoys employed thermionic valves in the electronic circuits, but the T9003 and the T1154 were later transistorised and were known as the T17053 and T17054.

They were 5 feet long, 9 inches in diameter, and weighed about 80 lbs, built to conform to the tripartite specification for C-size buoys. These two types of sonobuoy formed the sensor system of the Mark 1C Sonics System which was in service in the Royal Air Force in Shackleton and Nimrod MR1 aircraft until the early 1980's.

These buoys were very large and had to be carried in the aircrafts bomb buoy and so took space intended for torpedoes and depth charges.

It is of interest to note that unlike today these early UK sonobuoys were designed to be recoverable, refurbished and then reused whereas US sonobuoys were expendable.

T11514 on a Vibration Table in 1967 at the Foxbury
Electrical Installation Inspectorate *FAST collection*

# THE EMERGENCE OF LOFAR IN THE 1960S AND MUCH IMPROVED ASW DETECTION PERFORMANCE

In the early 1960s Low Frequency Analysis and Recording (LOFAR) technology was developed in the USA. This enabled particular low frequencies produced by submarine machinery and propellers to be detected rather than noise emitted over a broad band and initially led to extremely long detection ranges being feasible. The first LOFAR sonobuoy, the AN/SSQ-41 was fielded in 1965 by the USA as an omni directional passive, A-size sonobuoy. (36 inches long by 4.875 inches' diameter). The output of LOFAR sonobuoys was displayed on a frequency-time plot called a LOFARGRAM, an example of which is shown below.

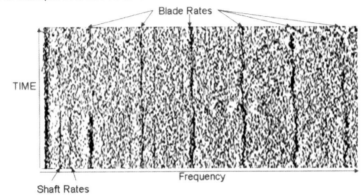

An example of a LOFARGRAM

Image credit: Defence Science and Technology Laboratory (Dstl), UK Ministry of Defence.

In 1968 the UK Ministry of Defence adopted the same technology to complement other British designs and a UK LOFAR began to take shape in late 1969 with the production of the omni directional A-size SSQ-48 (T24501) Jezebel sonobuoy. It was produced to a US Sparton Corporation design and had 31 radio channels. Ultimately 213,000 Jezebel sonobuoys were delivered by Dowty (now Ultra Sonar Systems) to the UK MOD.

In the Mk 1 Nimrod the LOFAR processing was carried out by the US Emerson AN/AQA5 processor. The physical output was provided by an electrical stylus scanning across moving paper, and the signal processing, including the spectral analysis, was done using analogue technology. However, with the availability of more capable digital technology, the MR 2 Nimrod carried out all the LOFAR processing digitally in the Marconi AQS 901 acoustic processor. The AQS 901 processor was followed shortly thereafter by the AQS 902 digital processor for the Sea King HAS Mk 5.

UK Jezebel  *Image UEL*

Many of the algorithms used in the digital processors were based on research carried out at the RAE.

Note the innovative use of a rotochute in place of a parachute for stability and retardation, whereas all buoys before and since used parachutes. Roger A Holler (the Ears of Air ASW) notes that the rotochute was used for many years, until the requirement for high speed high altitude launch forced a return to parachutes. LOFAR in the 1970s gave ASW aircraft exceptional detection performance until quieter submarines gradually came into service.

The design of LOFAR sonobuoys differed from the early broadband devices in that the cable from the flotation bag to the receive hydrophone was modified. A section of the cable of LOFAR devices was wound around a compliant rubber-based material, and the overall length of the cable was increased, with the aim of reducing the effect of surfaces waves etc. on the stability of the hydrophone, thereby reducing self-noise.

## THE COLD WAR PROVIDES THE IMPETUS TO INCREASE ASW SPENDING

The post-World War 2 1940s ASW community in the UK and the US realised that to make major strides in improving its ASW capability required a much greater understanding of underwater acoustics. Previously there had been significant progress in underwater acoustics research but particularly in the UK it had been a junior partner to research to support the development of radar in the 1930s and during World War 2.

In the early 1960s in the UK and US there was a further reason to pursue this course when the decision was made by both countries to adopt a nuclear deterrent based to a significant extent on ballistic missiles carried in Nuclear submarines. Developing acoustic detection systems capable of covert detection and reliable monitoring/shadowing of Warsaw Pact submarines became a major priority in the UK/US and a number of other NATO countries. All arms of NATOs ASW forces: submarines, ships and MPAs saw a steep rise in Research and Development (R and D) spend aimed at major improvements in acoustic sensor performance.

The RAE at Farnborough had become the MOD's centre of excellence for airborne ASW after world war 2 and benefited from this increase in R and D spend and was able to attract a number of top notch scientists to work in airborne ASW activities. As a result, the RAE was able to expand its airborne ASW activities and work in parallel on a number fronts as follows:

1. Conduct fundamental underwater acoustics research to better understand the underwater environment such that sonobuoys could be developed to exploit features favourable to long detection ranges.

2. Carry out applied research on all aspects of airborne ASW from the wet end buoy, the uplink to the MPA, the receivers and airborne processing and displays

3. Contract companies such as Ultra to produce prototypes etc. and to investigate their own novel ideas

4. Conduct trials with MPAs and friendly submarines to evaluate the RAE's ideas in the real world

5. Develop techniques and methods for predicting sonobuoy detection performance in any ocean area in which a MPA was likely to operate. Sonobuoy and sonar detection performance is far more variable over time and in different ocean areas than that of radar. Developing such techniques was an ongoing research topic for many years. The acoustic underwater environment is a very variable medium both in the short term minutes and hours and over the course of a day. As a result, it is difficult to model accurately in terms of its effect on sonobuoy detection performance.

6. Develop mathematical and digital models for predicting MPA mission effectiveness e.g. models to predict the probability of an MPA detecting submarine type xx transiting through a barrier of sonobuoys deployed across a submarine transit route.

7. Carry out the role of the UK MOD's trusted expert and advisor on airborne ASW and be the technical interface with the USA for exchange of information.

8. Call on the assistance of other MOD research establishments such as the AUWE at Portland and operational MPA crews.

Good relationships were built up with MPA crews who had experience of real operations and they provided valuable inputs to sonobuoy development activities.

The trusted expert role resulted in the specifications to which UK sonobuoys were developed being written by the RAE at Farnborough on behalf of the MOD. These specifications were written in the form of requirements such that industry were free to come up with their own hardware/ software solutions when bidding for development and production contracts.

Subsequently the RAE, carried out the assessment of the technical aspects of the resulting manufacturer's tenders for sonobuoy development contracts.

The AUWE at Portland provided inputs to the RAE on developments in the submarine and surface ship sonar scene and from its own sonobuoy research. Supporting research and trials were carried out at other MOD establishments to define the characteristics of potential target submarines in terms of noise signature, target strength, maximum diving depth and speed etc. to help guide the direction of sonobuoy development and for use in quantitative ASW performance assessment studies.

# CHAPTER 7

## NIMROD ENTERS SERVICE IN THE 1970S – THE WORLD'S FIRST JET POWERED MPA

On 4 June 1964, the British Government issued Air Staff Requirement 381 to replace the Avro Shackleton. Such a replacement was necessitated by the rapidly approaching fatigue life limits of the RAF's existing Shackleton fleet. A great deal of interest in the requirement was received from both British and foreign manufacturers: offered aircraft included the Lockheed P-3 Orion, the Breguet Atlantique, and derivatives of the Hawker Siddeley Trident, BAC One-Eleven, Vickers VC10 and de Havilland Comet. On 2 February 1965, British Prime Minister Harold Wilson announced the intention to order Hawker Siddeley's maritime patrol version of the Comet, the HS.801.

The Nimrod design was based on that of the Comet 4 civil airliner which had reached the end of its commercial life (the first two prototype Nimrods, XV147 & XV148 were built from two final unfinished Comet 4C airframes). The Comet's turbojet engines were replaced by Rolls-Royce Spey turbofans for better fuel efficiency, particularly at the low altitudes required for maritime patrol. Major fuselage changes were made, including an internal weapons bay, an extended nose for radar, a new tail with electronic warfare (ESM) sensors mounted in a bulky fairing, and a MAD (magnetic anomaly detector) boom. After the first flight in May 1967, the RAF ordered a total of 46 Nimrod MR1s and deliveries began in 1969.

Comets being Converted to Nimrod MR1 *(Photo: BAe)*

The Nimrod was the first jet-powered maritime patrol aircraft (MPA) to enter service. Aircraft in this role had commonly been propelled by piston or turboprop power plants instead to maximise fuel economy and enable maximum patrol time on station. Advantages of the

Nimrod's turbofan engines included greater speed and altitude capabilities, it was also more capable of evading detection methods by submarines, whereas propeller-driven aircraft are more detectable underwater by submarines with advanced sonars.

**Derrick McNeir of the RAE was the MOD's independent advisor on passive airborne acoustic processing for all versions of Nimrod, MR1, MR2 and MRA4. He reviewed the contractor's processing algorithms and then specified improvements.**

Nimrod MR1 *(Photo: BAe)*

In 1970 the Nimrod MR1 began entering service replacing the Shackleton:

Nimrod MR1s 1970 and a Shackleton *(Photo: BAe)*

The following Nimrod pictures and text are reproduced from the book BAe Nimrod written by John Chartres and published in 1986.

Nimrod MR1 Tactical Displays were a big step forward from the Shackleton and were based round Elliot Digital Computers. The Nimrod MR1's navigational functions were computerised, and were managed from a central tactical compartment housed in the forward cabin; various aircraft functions such as weapons control and information from sensors such as the large forward Doppler radar were displayed and controlled at the tactical station. The flight systems and autopilot could be directly controlled by navigator's stations in the tactical compartment, giving the navigator nearly complete aircraft control. The navigational systems comprised digital, analogue, and electro-mechanical elements; the computers were directly integrated with most of the Nimrod's guidance systems such as the air data computer, astrocompass, inertial guidance and Doppler radar. Navigation information could also be manually input by the operators.

Nimrod MR1 Tactical Displays

Nimrod MR1 Sonobuoy Store

Nimrod MR1 Sonobuoy Loading

Foxtrot Caught on the Surface by a Nimrod in the NE Atlantic

The design philosophy of these computerised systems was that of a 'man-machine partnership'; while on board computers performed much of the data sift and analysis processes, decisions and actions on the basis of that data remained in the operator's hands. After a mission, gathered information could be extracted for review purposes and for further analysis.

Coincidentally the prime contractor for the MR1's Tactical Weapons System was the author's previous employer, Easams Ltd, with a contract directly with the UK MOD and the development and testing rig was at their Lyon Way premises in Camberley.

The following is from an article in a RAF Historical Society Journal 33 written by Sqdn Ldr I M Coleman. The article describes Nimrod Operations in the Cold War.

> The Nimrod had a relatively high transit speed (400 kts, M0.69) and the ability to loiter as the fuel burned off, on three, or even two, engines that would give it a flexibility and speed of reaction much greater than its predecessor.
>
> The flight deck was 'Comet', with two pilots and a flight engineer. The first pilot might be the aircraft captain, but on most squadrons there would be about six pilot captains, two navigator captains and one AEO captain. A budding co-pilot could thus convert to first pilot with a back-end captain and get on top of his responsibilities for flying the aircraft before getting to grips with calling the tactical shots as well. The flying controls were pure 1950s too. They were powered by hydraulic servodynes from multiple hydraulic systems with much built-in redundancy, using a lethal traditional fluid. Certainly after flying home with a leaking wing system and a haze in the cabin, you had a headache for days.
>
> The shortened Cornet 4 fuselage had an underbody containing the radar scanner at the front and a full length, heated, unpressurised bomb bay. Rolls-Royce Spey engines replaced the Comets Avon's with some 'reaming out' of the wing root housing. The Spey proved to be very reliable and normally gave advance warning of problems. Given the engine location, asymmetric flying was almost an academic exercise.
>
> Aft of the flight deck was the toilet, forward door and then the two hemispherical beam lookout positions. With poor flight deck downward visibility, the ability to lean into the window and see down almost vertically was most useful looking for dinghies and the like. As the window distorted photographs taken through it, it was opened inwards and upwards to reveal optically correct fresh air. Being forward of the engines, the view was also unaffected by jet efflux, though one had to take care not to drop light meters down the intakes.
>
> Next came the Radio Operator's position. He had two HF radios and a LF receiver. Across the aisle were the Routine Navigator (Route Nav) and Tactical Navigator (Tac Nav) positions. The Route Nav operated the navigation equipment and carried out fixing with beacons, the radar, LORAN, Astro or (from 1982) Omega. On the Mk 1, the routine navigator's system was largely analogue, derived partly from the TSR2 project [authors note some of the Easams' designers of this system had worked on the TSR2]. The first generation inertial platform had to have what was called a 'run align'. One set the true heading of the runway on the box and selected it to 'run' as the aircraft rolled down the runway, giving it its heading

Nimrod MR1 and Shackleton *(Photo: BAe)*

reference. After a last minute runway change at Gibraltar, I can personally, vouch for the fact that it did not work well when set up backwards. The reversionary mode was the Doppler system and if that failed one could set in the estimated wind.

A most useful oddity was the Routine Dynamic Display (RDD). This projected an arrow, which was aligned with a chart taped to the table and gave an instant indication of the position and heading of the aircraft. By manipulating the illumination switch whilst running the arrow into a matchbox, the nav could convince gullible visitors that that was where he kept the arrow for safety!

The Tac Nav controlled the battle, usually initiating sonobuoy drops and managing the weapons. Feeding the large circular Tac Screen (supposedly the largest CRT of the time) was a new digital computer with 64K of memory. Including the two acoustic systems. this gave the aircraft a whacking 192K! The programme itself was run from a tape drive which you had to reload to access the Search and Rescue (SAR) version. One could designate markers and positions, receive data from all the sensors and throw it all into a Kallman filter to arrive at a target position to attack, whilst giving computer steers to the pilots' instruments. And it worked. Tactically, the drift of the system was noticeable, so one always homed to the radio signal of a related sonobuoy or to a smoke marker before attacking. The next generation inertial platform on the Mk 2 was far more accurate.

On the Mk 1 the Tac Nav would release the active and passive localisation sonobuoys of the Mk 1C system from the bomb bay. The 'Stage 2 Trainer'. inhabited by ancient aviators who passed the time growing copious quantities of tomatoes, could also simulate these from the ground. To drop a dummy buoy, you had to move a buoy indicator button to a vacant position on the store layout map for the bomb bay. You could get this wrong and drop a real buoy. As the Mk IC active buoys cost the same as an Austin Mini, this was not encouraged. The aircraft could be conned on to attack the target by the Tac Nav using either the computer algorithm or a manually plotted backup chart with the RDD. This was because on the Mk 1 the RDD was separate and unaffected by computer failure. On the Mk 2, the RDD went through the computer, so if this failed, the RDD did too! So much for progress.

Aft of the AEO on the starboard side were the positions for the Acoustics Co-coordinator and two operators (the 'Wet' team). On the Mk 1 a separate set was in place for the analogue Mk 1 C buoys. On the Mk 2, the new specialised attack buoys (CAMBS and Barra) were processed in the mainstream acoustic system. During a search, the three-man 'Wet' team would monitor sonobuoys. At the call of 'Action Stations' or 'Camera man up', two would leave and man the beam lookouts and take any hand-held photos needed.

The weapon loud could comprise elements from the old Mk 44 and newer Mk 46 American ASW torpedoes, with the British Stingray torpedo coming later on the Mk 2. There was also the standard Lindholme ASR gear, dinghy pairs, Containers Land Equipment (CLE), mail containers. 5-inch reconnaissance flares, explosive Anti-Submarine Target indicators [ASTI — replaced by the acoustic Signal Underwater Sound (SUS)] and smoke and flame floats (SFF). Harpoon missiles arrived during the Falklands War.

On notable occasions two 550 lb Special Weapons could be carried. Although the real things never came out of their store, practice 'shapes' were used for exercises. Once the source of great secrecy and a host of pedantic mandatory procedures, two of these practice Nuclear Depth Bombs NDB) are now on display at Hendon's RAF Museum.

## TRACKING SOVIET NUCLEAR SUBMARINES

The core Cold War activity for the Nimrod force was largely unknown to the public and to much of the rest of the RAF, it was to maintain surveillance of the Soviet submarine fleet. During the Cold War, the Soviets sent their nuclear submarines out into the Atlantic on a regular basis. In the event of a crisis, they would obviously be a danger but the assets to be protected at all costs were the SSBNs of our nuclear deterrent. Their patrol areas were highly classified and, as we had no need to know the details, we didn't. Positioned by SOSUS cueing, our task was to pick up the target and track it as covertly as possible, handing him on from aircraft to aircraft until handing over to other nations' MPAs or until other assets were in the trail.

This was done covertly to disguise as much as possible strengths and weaknesses in equipment capabilities and tracking techniques. Also, if it became apparent that aircraft tracking always began in particular areas, that would indicate areas of good SOSUS cueing from which it would be possible to build up a picture of the overall effectiveness, or otherwise, of that vital system. This cat and mouse game carried on throughout the Cold War.

The first aircraft laid a barrier of passive omni-directional Low Frequency Analysis and Recording (LOFAR) sonobuoys (the Jezebel system), which would pick up the submarine noise on the hydrophone and relay it by radio link to the aircraft. Most tracking was based on discrete frequencies produced by the power plant, machinery and generators of the target. These frequencies. or their harmonics, would pass as noise into the ocean through the hull. Later, with more computer processing a buoy that also gave a bearing (DIFAR) was used too. The buoy spacing was such that the submarine could not go through without being detected. One would update the water conditions by dropping a Bathythermal buoy, which dropped a thermometer on a line and radioed back the temperature profile. Based on this, one would decide the cable length to set on the sonobuoys.

Having detected the submarine on the barrier, it was then tracked by a series of sonobuoy patterns such as the live-buoy chevron. Once detected, an assessment of target position, course and speed could be made by getting more buoys in contact and comparing the received frequencies. This utilised the familiar Doppler shift principle of train whistles or racing cars. The frequency is higher when the target is coming towards you, is at the centre frequency as it passes (Closest Point of Approach or CPA) and is lower as it goes away. Once one had established the centre frequency, geometry allowed you to work out the angle the target was to a buoy. The speed was assessed from measuring known gearing or propeller blade readings or, after a full CPA had taken place, giving the maximum frequency shift against a formula.

If your assessment was good, he would CPA one pattern and, before he faded there, would fade in on the pattern ahead. On my last tracking sortie, I achieved the equivalent of a 'hole in one' when the submarine scraped along the buoy cable and cut off the hydrophone. On return I basked in some glory. My ever-tactful Lead Wet whispered in my ear that he would keep quiet about the fact that it was a wing buoy, not the pin buoy, that had been run down by the target! To hand over covertly involved leaving specific radio channel buoys indicating your assessment of the target. The off-going aircraft was required to be at least a stipulated distance down an outbound track from one of these buoys, whilst the incoming aircraft had to 'on-top' that buoy to tie it into his system and was not allowed to descend before a specific time and only along another specified vector. It was a rigid and very necessary procedure that has seen us safely through many operations.

Later in the era, we were sometimes instructed to carry out a 'passive attack' at the end of the sortie. A tight attack barrier or passive sonobuoys was dropped ahead of the target and the dropping of a weapon simulated. This involved over-flying the target on the attack run and in reasonable sea conditions, he would detect the over¬flight. To be within a button push of doing exactly what you would do in war, gave a huge feeling of achievement. Whilst other Cold War warriors studied target maps, we were actually up against our potential foe, day after day.

The basic Doppler tracking technique worked well for thirty years. By the end of the Cold War the targets were much quieter, but other techniques to exploit the sound in the ocean had been developed. Over the years the aircraft acoustic operators, the navigators and AEOs built up impressive levels of ability. In addition, the highly classified world of SOSUS produced some officers who gained an almost sixth sense for the patrol patterns of the Soviets. On several occasions, when devoid of reliable intelligence, these men have directed the aircraft to gain contact. Such skills are perishable and in the modern world with the occasions where they are needed sparse we are looking lack to a golden age of expertise.

## NIMROD MR2

Starting in 1975, 35 MR1 aircraft were upgraded to MR2 standard, being re-delivered from August 1979. The upgrade included extensive modernisation of the aircraft's electronic suite. Changes included the replacement of the obsolete ASV Mk 21 radar used by the Shackleton and Nimrod MR1 with the new EMI Searchwater radar, a new acoustic processor (GEC-Marconi AQS-901) capable of handling more modern sonobuoys, and electronic displays and

Nimrod MR2 *(Photo: Ultra)*

Nimrod Patrol painting by Patricia Forrest, GAvA *(Hansen Fine Arts)*

Nimrod MR2 *(Drawing: Hansen Fine Arts)*

not paper recorders as on the MR1 with a new mission data recorder and a new Electronic Support Measures (Yellow Gate) which included new pods on the wingtips. The MR2's new tactical computer provided a 50-fold increase in computing speed and power.

In 1981 this stunning painting of a Nimrod appeared on the cover of that year's Royal Airforce Yearbook, price seventy pence then.

A Victor refuelling a Nimrod *(Photo: BAE)*

Nimrod MR2 unpressurised carousel launcher *(Photo: UEL)*

Provision for in-flight refuelling was introduced during the Falklands War (as the MR2P), as well as hard points to allow the Nimrod to carry the AIM-9 Sidewinder missile to counter enemy Argentine Air Force maritime surveillance aircraft.

The Nimrod MR2 carried out three main roles – Anti-Submarine Warfare (ASW), Anti-Surface Unit Warfare (ASUW) and Search and Rescue (SAR). Its extended range enabled the crew to monitor maritime areas far to the north of Iceland and up to 4,000 km out into the Western Atlantic.

# UK'S HELICOPTERS ALSO DEPLOY SONOBUOYS

Formerly Sea King helicopters, and currently Merlin helicopters, based on surface ships and land bases, deploy sonobuoys for ASW search. Merlin is often used to localise and prosecute submarines detected by Sonar 2087, a long range active search sonar system fitted to the UK's ASW surface ships.

Merlin *(Painting: Neil Hipkiss, Aviation Artist)*

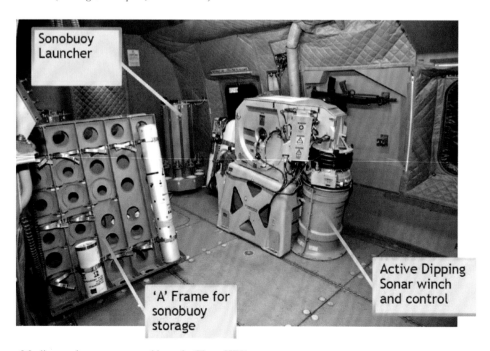

Merlin sonobuoy storage and launch *(Photo: UEL)*

# CHAPTER 8

## THE RAE'S SONOBUOY SYSTEM ACTIVITIES DURING THE HEIGHT OF THE COLD WAR IN THE 70S, 80S AND EARLY 90S

## INTRODUCTION

This chapter has been compiled from major inputs from former RAE Scientists and Alan Wignall and in particular:

- Les Ruskell – RAE/DRA
- Peter Viveash – RAE/DRA/DERA/QinetiQ
- Derrick McNeir – RAE/DRA/DERA/QINETIQ
- Ray Cyphus – RAE/DRA/DERA/DSTL
- Michael Buckingham – RAE/DRA/Professor in the USA where he now resides
- Trevor Kirby Smith – RAE/DRA/DERA
- Alan Wignall OBE - Ultra
- Peter Evans - AUWE

The chapter covers the following:

- Sonobuoy Size Reduction
- Channel – Selectable sonobuoys
- Super Barra Research at AUWE Portland
- Digital Sonobuoys
- Lithium Batteries
- GPS in Sonobuoys
- NATO Sonobuoy Specification
- Interoperability Trials
- International Collaboration
- RAE Trial's Capabilities
- Sonobuoys for Underwater Acoustics Research
- Sonobuoy Test Ranges

# SONOBUOY SIZE REDUCTION

In the 1970s Britain developed smaller lighter passive sonobuoys to allow larger numbers to be deployed during a search, designing in 1978, the world's first F-size sonobuoy, (304.800 X 123.825 mm) 12 inches long by 4.875-inch diameter. These were half the weight and one-third the length of conventional sonobuoys. Part of this reduction was achieved by replacing the internal air filled buoyancy chamber by a flotation bag inflated after water entry. The deployment of larger numbers of sonobuoys in a search field improved the chance of detecting hostile submarines.

The drive to shorten sonobuoy   length from standard A-size to F and G lengths (36", 12" and 16.5" respectively) was the UK's, with RAE and the RAF coming up with the idea and the technology driven by what was then Dowty (now Ultra). Dowty were able to design and produce shorter sonobuoys by using technology and less materials which had the same performance as the A-size versions and at a much lower cost per unit.

The US were in theory able to launch shorter buoys from their P3 aircraft by introducing a Sabot to fill the empty space in the SLC (Sonobuoy Launch Container). However, they didn't purchase any shorter sonobuoys for their P3s. In contrast the Canadians did use the shorter sonobuoys in their US made Orions which were essentially P3s.

Sonobuoys manufactured to this size, using a new tripartite 99-channel frequency allocation, were the SSQ 904 LOFAR and SSQ 906 wideband LOFAR. SSQ 937's. Bathythermal buoys were also produced in this size. Bathythermal buoys measured the temperature profile of the water they were deployed in and enabled MPA crews to evaluate the best depth at to deploy their search sonobuoys, UEL was the predominant producer of these smaller sonobuoys.

The method adopted by the US for launching sonobuoys was the Sonobuoy Launch Container (SLC). Buoys were loaded externally into the sonobuoy launch tubes, and ejected from the aircraft by a small pyrotechnic device.  The buoys are not accessible from inside the aircraft.  So the USN Lockheed P-3 Maritime Patrol Aircraft had external launch tubes.  By contrast, the UK did not adopt SLCs, but used 'bare-buoy' launchers inside the cabin. The UK Nimrod MR1 and MR2 used two rotary carousel launchers, each capable of holding up to six buoys and two single buoy launchers. The UK Merlin aircraft also uses two unpressurised carousels, but of 10 buoys each. (The Nimrod MRA4 would have used similar carousel launchers to those on the Merlin). To use the carousel launchers, the aircraft cabin had to be de-pressurised, effectively limiting its use to fairly low altitude (not generally a disadvantage); the single launchers were pressurised, giving the option of launching buoys from high altitude.  All the launchers could be reloaded in flight.

The use of internal launchers had the advantage that a large stock of sonobuoys could be carried within the Nimrod cabin on storage racks.  As more diverse buoy types became available in the late 70s and early 80s, internal carriage helped obviate the need to make decisions on buoy mixes before the start of the mission; buoys could readily be removed from the internal launcher and swapped with other types if required.  Furthermore, the move from 31 to 99 RF channels combined with internal launchers meant that 'blocked' RF channels could more readily be accommodated.

The use of internal launchers meant that there were significant advantages in reducing the length of the buoys. The move from A-size Jezebel to F-size (i.e. from 36 inches in length to 12 inches in length) pioneered by UEL meant that three times as many passive omnidirectional buoys could be carried in the aircraft's storage racks.

The success of the F-size led, in the mid-1980s, to the development of the G-size DIFAR buoy, 16.5 inches in length. It was not possible to accommodate the more complex sensor and associated electronics within the F-size, but the G-size became the standard for UK DIFAR. The Bathythermal buoy was the only other buoy to be reduced to F-size. Barra and CAMBS remained at A-size because of their greater complexity.

## CHANNEL-SELECTABLE SONOBUOYS

Buoys of the 1970s were all on one of 31 channels, with the transmitter frequency determined by a crystal-controlled oscillator. As submarines began to get quieter and more difficult to detect, the limitations of 31 VHF channels started to be felt, particularly where a number of ASW platforms were working in the same area, or when working near the littoral where interference could be experienced from civil radio traffic such as taxis. Although buoys could be set to cease RF transmission after one hour, thereby freeing up an RF channel, often buoys were set with longer lives and channels could remain 'blocked' during a whole sortie. The limit of 31 channels therefore remained a significant constraint. The RAE and the RAF and MOD agreed that extra channels were required.

In 1972 agreement was reached at the classified 1972 CCITT (now the ITU) World Administrative Radio Conference (WARC), Geneva, to extend the sonobuoy band from 31 channels (between 162.25 and 173.5 MHz) to 99 channels (between 136 and 173.5 MHz). The 99 RF channels are equally spaced at intervals of 375 kHz, between 136 and 173.5 MHz, except for missing out the frequencies 161.5 and 161.875 MHz, which would, theoretically, be sonobuoy channels '100' and '101'. It is unclear why that gap, which lies at the top end of the VHF Marine Mobile Band, was left in the sonobuoy RF band. However, conveniently, the Automatic Identification System (AIS) now uses frequencies in that gap, which avoids mutual interference, and enables sonobuoy receivers to receive AIS signals

The 99-channel VHF FM telemetry scheme remains in use to this day, suitably adapted for transmission of digitised acoustic and other data. However, the trend towards operation in littoral waters, or at high altitude, presented particular difficulties for sonobuoy reception, due to interference from land-based and coastal VHF transmissions. Coping with RF interference became a key research topic for the UK and for the US, from the late 1990s, which led to a new robust and more versatile digital telemetry standard, based on the same RF channels, but using sophisticated modulation and encoding techniques, which is now known as STANAG 4718, and which is in the process of being adopted and ratified at the time of writing this book.

The sonobuoy receivers in the Nimrod MK1 were only capable of receiving 31 channels, whereas the MR2 was capable of receiving the full 99 channels.

Whilst the increase to 99 channels was a major operational step forward, taking advantage of the increase in the number of channels gave rise to a logistical problem, in that stocks of more buoy channel variants had to be held. Buoy transmitters originally had their frequencies fixed by choice of crystal at manufacture, and could not be changed once manufacture was complete. But the emergence of cheap microprocessors in the early 1980s meant that phase-locked loop frequency control could cheaply be incorporated to programme buoy transmitters so that channels could be selected immediately prior to launch. Agreement was reached within the western sonobuoy-manufacturing nations on an interface between aircraft and buoy which would enable channels to be selected during a mission by the aircraft mission system. Because cost is such an important driver for sonobuoys the buoy interface had to be extremely simple. The aircraft 1553 data bus was therefore interfaced to a dedicated control unit in the sonobuoy launcher. Communication between the control unit interface and the buoy itself was based on self-clocking Manchester code and simple spring-loaded electrical contacts. The RAE was involved in airborne trials to assess the integrity of the proposed interface, a prototype version of which was built by UEL.

A Remote Function Selection (RFS) standard was defined, actually based on infra-red communication between the buoy and a hand-held remote control unit, or between the buoy and the launcher, rather than using electrical contacts, and was incorporated into both US and UK sonobuoy generic specifications. However, it was removed from the US and UK specifications in 2000 and 2006 respectively, and was never implemented in a production buoy. An adequate solution, using manual selection of operating parameters (RF channel, life, depth) using buttons and an LED display on the buoy, was defined and implemented on all sonobuoys, known in the UK as AFS (Autonomous Function Selection) and in the US as EFS (Electronic Function Selector). The US also implemented a Command Function Selection (CFS) scheme on all buoys, active and passive, which can change buoy parameters, including RF channel, after launch, using the UHF RF downlink. The UK chose only to implement a command downlink on its active buoys (CAMBS SSQ963 and ALFEA SSQ926), and not on its passive buoys.

## SUPER BARRA RESEARCH AT AUWE PORTLAND

"Super Barra" was a horizontal planar array, based on the Australian SSQ801 buoy, named 'Barra', which later evolved to become the UK SSQ981E Barra buoy.

Peter Evans joined the airborne sonar team at Portland in 1979, working with Dennis Stansfield, Alan Lanham and Willie Wildash on the applicability of large aperture sonobuoy arrays for passive submarine detection. By the time he arrived, the team had already instigated a number of studies in industry/academia to assess the feasibility of a "super Barra" for improved detection of quiet submarines. At that time, sonobuoy work in the UK was primarily split between Plessey at Templecombe, and Dowty at Greenford, and MOD PE had the rather enlightened policy of trying to keep both in business by splitting production between them. As a result, both companies were made aware of the outcome of the research, and some of the research eventually found its way into the 903 studies.

In parallel with the array design work, the team developed a Modular array to evaluate the benefits of large apertures in real environment. The array was constructed from aluminium tube, and deployed from a large surface float. The design was flexible, and enabled array apertures of up to 50ft to be evaluated at sea. A key part of the investigation was an assessment of noise directionality (or anisotropy) at low frequencies, the aim being to determine how big arrays needed to be to provide the required performance improvements. A key part of the work was an investigation into the use of Adaptive Beamforming to reject interference from ships, thereby enabling detection of quiet submarines in the presence of noisier targets. This work was carried out by Alan Smith, who worked closely with RSRE Malvern to design and apply the algorithms. A series of trials were carried out, including participation in the "Danny Boy" trials, run by TTCP. A lot of useful work was carried out as part of the programme, including the development of improved methods for decoupling the array from surface motion.

# DIGITAL SONOBUOYS

One of most important innovations in UK sonobuoy design initiated by the RAE Farnborough since World War 2 was the move from analogue to digital electronics in UK sonobuoys. Peter Viveash was the RAE champion for this innovation which resulted in HIDAR a dual mode analogue digital version of DIFAR and the UK's first digital sonobuoy.

The key advantage of digital sonobuoys is better detection performance in noisy sea environment HIDAR was planned to be the RAFs and RNs main passive sonobuoy from the late 90s onwards replacing analogue DIFAR and the development and production contract was placed after a competitive tender.

DRA/DERA had a major role in this tender as follows:

- Wrote the technical specification for HIDAR
- Devised a quantitative tender assessment marking scheme for the technical aspects of HIDAR
- Attended MOD/Competing Contractor briefing meetings as MOD's technical advisor
- Carried out the tender assessment of the technical aspects of the bids and documented the outcome
- Attended the MOD's Tender assessment meetings which resulted in Ultra Electronics winning the competition

MOD wrote a thank you letter to the DERA as follows

**Directorate of Maritime Projects**
Procurement Executive, Walnut 0c, MOD Abbey Wood #73,
BRISTOL, BS34 8JH

Telephone: 01179134769
Fax: 01179134956

---

| | |
|---|---|
| Dr R N Andrew | Your Reference : |
| General Manager | |
| Sensor & Avionics Systems Dept | Our Reference : D/RMPA/81/2 |
| Air Systems Sector | |
| DERA Farnborough | Date : 19 December 1997 |
| HANTS GU14 6TD | |

---

Dear Neil,

## HIDAR SONOBUOY ASSESSMENT

The MOD announced this morning that it intends to place a contract with Ultra Electronics Ltd for the HIDAR sonobuoy. This milestone is a significant achievement and represents the culmination of many years of work by the Maritime Systems Group. From our standpoint, we recognise that DERA's work on HIDAR is an excellent example of how we can achieve technology "pull-through" from the research programme to industry.

UEL were, of course, selected through competition and I would like to thank you for DERA's extremely professional technical assessment of the bids. In particular, I should be grateful if you would pass on our appreciation to Mark Brown, Peter Viveash, Peter Martinson, Mike Ralph and Clive Radley. We now look forward to working with them to ensure that the contractor delivers the goods!

Wishing you a Happy Christmas,

yours sincerely,

When HIDAR was first delivered to the RAF, the acoustic processor the AQS 971 on the Nimrod MR2 hadn't yet been modified to process HIDAR digital mode. To enable the customer to get an early measure of HIDARs digital performance compared with the in service analogue DIFARs DERA planned and took part in a comparison trial. DERA was allocated 2 Nimrod sorties in an already scheduled RAF trial at AUTEC whose main aim was to evaluate the AQS 971. DERA developed digital HIDAR processing and displays on a PC based partial 971 emulators which was used during its two sorties to process the uplink information from co located DIFARs and HIDARs.

At ISIC 2002 a presentation on behalf of the DPA reported:

In-service experience so far indicates greater reliability and improved performance of digital HIDAR over DIFAR. Trials in digital mode have demonstrated significantly improved performance (over standard DIFAR) particularly in noisy environments. These trials were planned and analysed by DERA with the assistance of Nimrod crews and Ultra.

The following is from an Ultra Electronics presentation describing their implementation of digital electronics.

Surface Unit:

- Analogue: Signal processing by voltage manipulation using electronic circuits, e.g. op-amps
- Signal processing limited to simple functions
- Processing variation between buoys
- Digital: Software-based signal processing
- Consistent software-based processing
- Complex functions possible

Telemetry Link:

- Analogue: Modulation of the signal voltage onto a carrier
- Need tight control of component tolerances
- Digital: Direct digital synthesis of RF waveform
- High precision digitally generated waveform
- More sophisticated modulation schemes achievable

## LITHIUM BATTERIES

A, F and G-size passive sonobuoys traditionally employed magnesium-lead-chloride or magnesium-silver-chloride seawater batteries. These are inert until flooded, do not deteriorate in storage with age as do more conventional batteries, and are relatively cheap. The main driver of battery capacity is the need to provide 1 watt of RF power for the life of the buoy. However, the CAMBS active sonobuoy needs a much higher capacity battery, with the requirement driven by the active transducer. The battery in the initial CAMBS design was a silver chloride seawater battery and was very expensive. There was therefore a requirement for a cheaper, high capacity battery with a long shelf life, such as the lithium batteries that were starting to become available.

High performance lithium batteries are now commonplace, particularly in everyday electronic devices such as laptops, tablets and mobile phones. However, in the mid- 1980s they were relatively novel. The high capacity lithium-sulphur dioxide cells identified as suitable for use in CAMBS contained enough pressurised liquid sulphur dioxide in one D-size cell to incapacitate the crew of an aircraft should the cell vent. Accidental short-circuiting of the cell could cause an explosion. Circumstances which would cause battery malfunction also often led to distortion of the cell and buoy casing, making it impossible to dispose of a venting buoy through the sonobuoy launch tubes. Alternative high capacity lithium technologies, such as lithium-thionyl chloride, posed other problems.

# GPS IN SONOBUOYS

In the localisation and torpedo release phase of an MPA ASW sonobuoy based operation, accurate knowledge of the target's position is vital to maximise the probability of sinking the submarine. This in turn requires accurate knowledge of the localisation sonobuoys' positions.

GPS is well-known for its military uses and was first developed by the US to aid in its global intelligence efforts at the height of the Cold War. The GPS or Global Positioning System comprises a constellation of orbiting satellites. By measuring the time difference of arrival of signals from several satellites using a GPS receiver and from knowledge of the satellite orbits the location of the receiver can be ascertained accurately.

The RAE's Trevor Kirby Smith led an extensive research programme in the 1990s investigating the provision of GPS on sonobuoys. Before the GPS era the techniques for localising sonobuoys once deployed were relatively inaccurate, time-consuming and involved aircraft manoeuvres in the vicinity of the buoy. One strand of this research looked at Differential-GPS as an alternative to manoeuvring.

The section below is the introduction to the paper.

Brown, Peter, Kirby-Smith, Trevor, "Operational Field Trials of GPS Equipped Sonobuoys," *Proceedings of the 9th International Technical Meeting of the Satellite Division of the Institute of Navigation (ION GPS 1996)*, Kansas City, MO, September 1996, pp. 1553-1561.

*This paper describes a trial involving the NAVSYS S TIDGETTM low-cost GPS sensor in fully-functioning sonobuoys in an operational environment. Sonobuoys are disposable air-launched buoys that are used in Anti-Submarine Warfare (ASW). Current techniques for localising the buoys once deployed are relatively inaccurate, time-consuming and involve aircraft manoeuvres in the vicinity of the buoy. Differential-GPS has been proposed as an alternative to the present localising techniques, however the sonobuoy environment leads to problems with a conventional GPS implementation. - Sonobuoys are disposable, and are therefore extremely cost sensitive. - Sonobuoys are stored for extended periods (many years), but must be operational immediately when deployed. - Large thermal transients occur on deployment from the aircraft into the sea. - The GPS antenna is subject to a high degree of masking due to its low elevation above the sea surface, aggravated by high sea-states and/or high winds. - Sonobuoys are battery powered and have limited available spare power for a GPS receiver. - Sonobuoys broadcast 1 Watt of RF power adjacent to the GPS antenna. This paper describes the GPS technology used for this application and its implementation in a fully-functional sonobuoy, and presents the results from a trials sortie flown in October 1995 using an RAF Nimrod Maritime Patrol aircraft and 9 sonobuoys in the water at the NATO AUTEC ASW range in the Bahamas. The results presented include a comparison between the TIDGET positional accuracies and the range instrumentation, and a summary of objectives met.*

It is interesting to note the concerns about cost bearing in mind what has happened since then with GPS.

At the time, around 1990, both the cost and size of available GPS processing chipsets capable of receiving the GPS coarse signal prohibited their use in sonobuoys. The point of TIDGET, which was marketed by a small firm Navsys was that the buoys would receive the GPS signal and relay it to the aircraft after minimal processing. Essentially the main processing of the GPS coarse code was done in the aircraft, and the need for cheap and miniaturised chipsets was thus largely removed.

Trials of the Navsys 'Tidget' 'decimate-and-relay' GPS technology, in comparison to 'full engine' GPS Modules, took place up to 1997, under the auspices of DRA(F).

Ultra's HIDAR proposal in 1997 included 'for but not with' provision for GPS, without commitment either to the 'decimate-and-relay' or 'full engine' solutions.

However, from 1997 there was a decisive move in thinking away from Decimate and Relay (D&R) in favour of Full Engine (FE).

Some key factors which drove that decision were as follows:

- The USA removed selective availability, thereby making standard GPS sufficiently accurate.
- The advent of low-cost multi-channel 'all-in-view' correlators made it possible to get a rapid time-to-first-fix from a cold start.
- Wider supplier base for GPS Modules.
- Ease of standardisation of GPS data
- An obvious technology miniaturaisation and cost reduction trend for GPS modules.
- Much simpler airborne installation.

There was little technical risk, and it was straightforward to include GPS data in the UK telemetry format, which was already digital.

Therefore, the technology progressed directly to a production contract without further study.

## NATO SONOBUOY SPECIFICATIONS

The sonobuoy-manufacturing and sonobuoy-using nations within NATO have agreed a set of standards, enshrined in a document known as the NATO Sonobuoy Specification, which were designed to enable nations within NATO to launch and process each other's sonobuoys. This document sets standards for buoy form and fit and RF and telemetry characteristics. The sonobuoy community was always held up by NATO as the shining example of what could be achieved if interoperability were taken seriously. Admittedly the issues with sonobuoys were rather simpler to resolve than, for example, with complex weapons systems, but there are undoubtedly other military systems in use where NATO nations could have used one another's stores had thought been given at an early stage to interoperability.

The NATO Underwater Warfare Capability Group, Maritime Air Syndicate (UWW-CG-MA), which oversees specialist teams (STs) and an Industry Support Team (IST), continues to this day, with the development and sharing of interoperable standards and joint capabilities being a key part of its role.

## INTEROPERABILITY TRIALS

It is, of course, one thing to claim compliance with an interoperability specification, quite another to demonstrate that NATO nations can, indeed, deploy and process each other's sonobuoys. To test this a number interoperability trials have been conducted.

The first was held at Nimes Garons airfield in France. This involved MPAs from the UK, US, Canada and France and Germany not all nations had the capability to process all buoys. For example, the UK could not process the US DICASS (SSQ62) and the US could not process the UK CAMBS (SSQ963). However, those buoys which could in principle be processed were exchanged between nations. Each nation then conducted a trial to deploy and process the exchanged buoys and wrote a report on their experiences. The reports included, for example, experiences of buoy handling, loading, function selection, deployment, telemetry and processing. Where possible, any issues or difficulties were subsequently addressed.

Because a lot of changes had taken place to sonobuoy hardware technology during the 1980s, a further interoperability trail was held. Instead of MPAs meeting in one place, a programme of mutual buoy exchanges was agreed between the participating nations. Packages of buoys were then physically exchanged according to the agreed schedule and handling and deployment trials conducted independently by the participating nations. Again, results of trials were recorded and written up. The report of the outcome of these trials became, in effect, the evidence of interoperability in practice.

## RAE TRIALS CAPABILITIES

RAE had the capability to conduct extensive air and sea trials of new sonobuoy designs. At the RAE Farnborough site an anechoically clad acoustic test tank was available. However, the principal trials asset was the RMAS Colonel Templer. This was a converted ocean-going stern trawler, launched in Hull in 1966 as the 'Criscilla'. Displacing 1,300 tons and 180 foot in length, the Criscilla was refitted as a sonar laboratory and named after Colonel James Templer, the first Director of the Royal Balloon Factory, the precursor to the Royal Aircraft Factory and later the RAE.

Colonel Templer     *Les Ruskell Picture*

Importantly, it was equipped with bow thrusters to aid manoeuvrability, acoustically isolated generators to provide electrical power for the lab when the main engines were switched off to reduce noise, a platform low down in the side of the ship to enable buoys to be launched and recovered, and a capability to deploy and tow underwater sound sources to simulate submarines. The ship was based in Newhaven, but during the trials season usually worked out of Falmouth or Stornoway, often spending several weeks at a time at sea. During the 1980s it also conducted trials further afield in locations such as the Azores and at the AUTEC Range in Bermuda.

The RAE also had two aircraft equipped for launching sonobuoys and recording and processing their signals. BAC 1-11, XX 919, had two single-tube launchers, and Andover, XS 607 had four launchers activated by a clockwork mechanism to enable sticks of buoys to be deployed. Both could have racks of roll-on-roll off equipment fitted containing sonobuoy receivers, instrumentation tape recorders and an AQS 902 Sea King Sonics Processor and display unit.

BAC 111    *Les Ruskell Picture*

The combination of a well-equipped sonar laboratory on the Colonel Templer, two aircraft capable of monitoring, recording and processing sonobuoy signals both in conjunction with or independent of the ship, and communications equipment enabling operation also with MPA, ASW helicopters and, on occasions, UK submarines, provided the RAE with a very powerful capability for the conduct of ASW research and development.

Andover     *Les Ruskell Picture*

The picture and list below is of the team and itinerary from a February 1994 research trial, which was part of a Project Definition study for a Future Passive Surveillance System (project name SR(SA)902). The team comprised a mix of staff from industry and from DRA Farnborough. The late Mike Ralph (2nd row, 2nd from right), who was Chief Engineer at GEC Marconi Naval Systems Templecombe (formerly Plessey), subsequently worked for various MOD research organisations, and at DSTL became a champion for miniaturised sonobuoy systems for unmanned airborne and distributed ASW, which, at the time of writing, is an active research and development topic.

Colonel Templer example research trials team, Madeira / Morocco, February 1994 - *UEL picture*

Back row: Steve Broadmeadow DRA(F) 'SNR Station', Simon Grosvenor Ultra 'Floating Resource', Wayne Terry GMNS 'Electronics'

Middle row: Andy Armstrong '1st officer', Brian Maclaren DRA(F) 'Sound Sources', Steve Pagan GMAv 'SNR Station', Ken Clayton DRA(F) 'Trials Director', Crew, Adam Warren DRA(F) 'Sound Sources', Mike Ralph GMNS 'VLA Trials Leader', Dave Hillier DRA(F) 'Processors'

Front row: Alan Wignall Ultra 'Trials Technical Controller', Brian Holden DRA(F) 'Tape Recording', Crew, Crew, Ernie 'Bosun', Tex 'Crane Operator'

GMAv: GEC-Marconi Avionics (Rochester)

GMNS: GEC-Marconi Naval Systems (Templecombe)

SR(SA)902 STUDY - DATA COLLECTION TRIALS - ITINERARY

| | |
|---|---|
| Monday 17 Jan 94 | CT sails from Portsmouth |
| Sunday 23 Jan 94 | CT arrives in Porto Santo |
| Tuesday 25 Jan 94 | Rest of team fly on BAC111 to join CT |
| Wednesday 26 Jan 94 | Lab setup and equipment shakedown |
| Thursday 27 Jan 94 | **HARP shakedown, followed by trial CT1**<br>Madiera deep, Static sources |
| Friday 28 Jan 94 | VLA shakedown |
| Saturday 29 Jan 94 | VLA trial S1 |
| Sunday 30 Jan 94 | **HARP trial CT1A** (full rerun of CT1)<br>Madiera deep, Static sources |
| Monday 31 Jan 94 | Consolidation day |
| Tuesday 1 Feb 94 | **VLA trial S1A** (rerun of S1)<br>Madiera deep, Static sources |
| Wednesday 2 Feb 94 | Consolidation day |
| Thursday 3 Feb 94 | Madiera. Spare parts for crane |
| Friday 4 Feb 94 | Dip clear hiccup; Head for Gibralter |
| Saturday 5 Feb 94 | In transit to Gibralter |
| Sunday 6 Feb 94 | In transit / arrive at Gibralter |
| Monday 7 Feb 94 | Clearance received - Depart for Morroco |
| Tuesday 8 Feb 94 | In transit to Morroco |
| Wednesday 9 Feb 94 | Arrive at trial site. **VLA trial S2**<br>Morroco deep, Static sources |
| Thursday 10 Feb 94 | **VLA trial D1**<br>Morroco deep, Moving sources |
| Friday 11 Feb 94 | **HARP trial CT3**<br>Morroco shallow, Static sources |
| Saturday 12 Feb 94 | **HARP trial CT4**<br>Morroco shallow, Moving sources |
| Sunday 13 Feb 94 | Rest day - Lanzarote |
| Monday 14 Feb | Trials team return to UK on BAC111 |

# SONOBUOYS FOR UNDERWATER ACOUSTICS RESEARCH

During the Cold War the UK MOD invested heavily in fundamental acoustics research at Admiralty Research Establishments in order to better understand how submarines could hide in the ocean and to better predict the performance of sonar systems. Radio and Navigation Department at the RAE Farnborough contributed significantly to this research, in particular making use of the unique capabilities of air-deployed sensors. To this end a number of special-purpose research sonobuoy types were commissioned by the RAE for more fundamental research into underwater acoustics, and a couple of examples are described below.

At the beginning of the 1980s a batch of Vertical Line Array (VLA) sonobuoys was procured from UEL. These buoys each deployed an array of hydrophones vertically, with the output from each calibrated hydrophone being multiplexed into a standard sonobuoy RF channel and available at the monitoring receiver after de-multiplexing. The purpose of these buoys was to study acoustic propagation on the continental shelf around the UK at frequencies where the acoustic wavelength is comparable to the water depth and Modal propagation predominates.

A series of VLA sonobuoy experiments was conducted in the Southwest Approaches to the English Channel in 1981 as part of a multinational sensor trial known as "Danny Boy", involving the coordinated use of fixed-wing aircraft, helicopters and surface ships. These experiments were followed by a major series of trials in 1984 in which the VLA sonobuoys were air-deployed at a number of locations within the waters over the continental shelf around the British Isles (including the Northwest Approaches, the Southwest Approaches and the North Sea). The RAE's Research Vessel Colonel Templer was used to tow underwater sound sources at different ranges from the sonobuoys in order to measure the shallow-water acoustic propagation loss in different shallow-water environments. A series of trials was also undertaken to obtain improved seasonal measurements of the ambient noise at various deep-water locations around the UK.

Accurate measurements of the background ambient noise in the ocean are needed to calculate the detection ranges of sonar systems, and at the frequencies of most interest previous measurements of ambient noise using sonobuoys were known to be contaminated with hydrophone self-noise and RF transmitter noise. Specially calibrated sonobuoys were procured from UEL and deployed from the RAE BAC 1-11 aircraft at five deep-water locations off the UK continental shelf at three-monthly intervals. These experimental sonobuoys were designed specifically for the purpose of measurement over a limited frequency range to overcome the above problems, and were carefully calibrated and designed to reduce self-noise.

A very large number of acoustics research papers were also published by the RAE in the 70s and 80s but after the USSR collapsed the funding for ASW was reduced and subsequently the RAE produced few pure research papers.

# SONOBUOY TEST RANGES

The MOD maintained a number of ranges which could be used in sonar/sonobuoy work. The submarine noise testing range at Loch Goil was frequently used by RAE and sonobuoy manufacturers to test and calibrate equipment, having the advantage of low ambient noise levels. Routine testing of sonobuoys off the production line was carried out at RAE West Freugh, near Stranraer. In accordance with BS 6001-1/ISO 2859-1 (Sampling Procedures), samples were taken from each manufacturing batch.

In accordance with the appropriate British Standard] on sample testing, samples were taken from each manufacturing batch of sonobuoys and taken to the range at West Freugh. If all buoys in the batch worked satisfactorily the whole batch would be accepted into service by MOD. If there were failures, then a further sampling took place. If no failures were experienced then the batch would be accepted, but further failures would require the batch to be investigated and possibly re-worked. Later this facility was moved to BUTEC and was managed by QinetiQ.

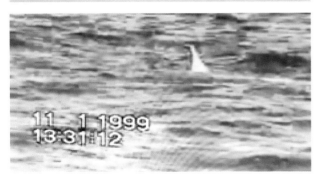

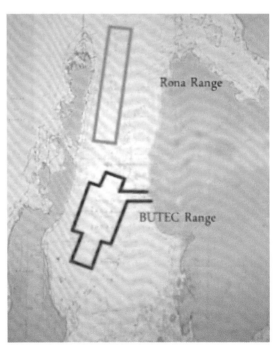

1999 Barra Deployment Trial at West Freugh

BUTEC range    *UEL picture*

# INTERNATIONAL COLLABORATION

The agreement between a number of countries re sonobuoy standardisation led to an International Sonobuoy Interoperability Conference (ISIC) being set up in the 1980s and held in various NATO countries every two years, under the auspices of NATO maritime Capability Group 4 (MCG4), now reorganised as NATO UWW-CG-MA.

The UK and the USA were the prime movers behind this idea. ISIC was intended to ensure new generations of sonobuoys were designed with a requirement that they could be deployed from all of the participating countries' airborne ASW assets. The conference also served as a platform for exchanging information on the latest developments in sonobuoys and related technology. Although loosely based around NATO, participants also initially included Australia and New Zealand as major sonobuoy users (and in the case of Australia, manufacturer). An essential element of these conferences was that industry were also invited, and were able to discuss issues of common interest in the manufacture and supply of sonobuoys. The format was for the Government representatives to make a keynote statement of their nations' vision of where they were going in terms of sonobuoy development (for example, EFS, RFS, size reduction), and identify where these might lead to interoperability issues. Sonobuoy manufacturers were then able to make presentations on interoperability issues as they affected industry.

The conference was attended by representatives from the various MOD's, sonobuoy manufacturers, and MPA aircraft ASW tasked crew members. Peter Viveash of the DRA chaired a number of the early conferences.

ISIC conference locations are listed below:

|        | Year | Location          | Theme                                                    |
|--------|------|-------------------|----------------------------------------------------------|
| 1st    | 1980 | Warminster, US    |                                                          |
| 2nd    | 1981 | St. Raphael, FR   |                                                          |
| 3rd    | 1984 | Ottawa, CA        |                                                          |
| 4th    | ?    | ?                 |                                                          |
| 5th    | 1992 | Koblenz, GE       |                                                          |
| 6th    | 1994 | Livorno, IT       | Interoperability, the Sonobuoy System and the Environment |
| 7th    | 1996 | Toulon, FR        | Blue, Brown and a Little Bit of Green                    |
| 8th    | 1998 | Bristol, UK       | A L'écoute De La Mer (Listening to the Sea)              |
| 9th    | 2000 | Patuxent River, US | The Future is Active                                     |
| 10th   | 2002 | Halifax, CA       | Many Listeners Hear More                                 |
| 11th   | 2004 | Toulon, FR        | Do it Digital                                            |
| 12th   | 2006 | Berlin, GE        | Green and Clean Comms Machine                            |
| 13th   | 2008 | Washington, US    | A Recall to Life for Sonobuoys                           |
| 14th   | 2011 | Halifax, CA       | Doing More with Less                                     |
|        | 2014 | -- Not held --    |                                                          |
| NUWIC  | 2016 | La Spezia, IT     | Shipborne, Airborne and Underwater Multistatics          |

The 2002 ISIC programme is shown below. The theme was multistatics.

# 10th International Sonobuoy Interoperability Conference (ISIC) 2002

## " *Many Listeners Hear More* "

### AGENDA and ABSTRACTS

## 23-25 October 2002

## Halifax, Nova Scotia, CANADA

## GENERAL INFORMATION

The 10th International Sonobuoy Interoperability Conference (ISIC) will be held in conjunction with the meeting of NATO NG4/SG41. The meetings will be held in Canada at the Casino Nova Scotia Hotel in Halifax, Nova Scotia, from 23 – 25 October 2002.

The theme of the ISIC 2002, "Many Listeners Hear More", has two interpretations. Many listeners could refer to multistatics and the ability to work more effectively using this concept. A second interpretation could be that multiple nations working together on the ASW problem provides a better solution, a reference to interoperability and the NATO alliance.

A site office, located in Suite 288, has been established at the hotel on the same floor as the meeting room. Registration will occur in front of Nova Scotia D Ballroom on Tuesday and Wednesday morning.

The agenda, social events and abstracts of presentations are listed on the following pages. Exhibits will be on display in Acadia Room C, which is located on the main floor of the hotel directly below the ISIC site office.

There will be morning and afternoon breaks each day for conference attendees. Lunch will not be provided.

If you have any questions or require assistance with anything, please see Karen Isenor in the ISIC office. The office will be open during the hours of the conference for your convenience and we have the ability to fax and copy small quantities.

Welcome to Canada and we hope you have a pleasant stay.

## AGENDA – OCTOBER 23

| Time | Topic | Presenter |
|---|---|---|
| 0730 - 0830 | Registration & Coffee | Outside Nova Scotia D |
| 0830 - 0850 | Welcome / Admin / Intro | Maj Sullivan / LCol Poulin (DAEPM(M)) |
| 0850 - 0910 | Opening Remarks | Guest Spkr Colonel Hache MAC(A) |
| | *National Presentations* | |
| 0910 - 0930 | UK | Clive Radley |
| 0940 - 0950 | US | Peter Verburgt |
| 0950 - 1010 | Canada | Maj Sullivan |
| 1010 - 1040 | COFFEE | Outside Acadia C |
| 1040 - 1100 | France | Laurent Gaillard |
| 1100 - 1120 | Germany | Alex Both |
| 1120 - 1140 | Australia | Richard Wolf |
| 1140 - 1200 | Italy | Ghibaudi |
| 1200 - 1220 | Turkey | LCdr Yalcin |
| 1220 - 1350 | LUNCH | On Your Own |
| 1350 - 1520 | Detection of FIAC | QiniteQ |
| 1520 - 1550 | In-Buoy Signal Processing | USSI |
| 1550 - 1620 | COFFEE | Outside Acadia C |
| 1620 - 1650 | Continuous Active Sonar | S&CS |
| 1650 - 1720 | Network Centric Warfare | Thales |
| 1800 - 1900 | Group Photo | Location to be determined – in the hotel |
| 1900 - 2100 | Province of Nova Scotia Reception | Halifax Ballroom A in the Hotel |

Programme for the 2002 ISIC held in Canada

ISIC 2002 attendees, including many R&D leaders from government and industry across NATO (and Australia). The chairman Peter Verburgt is pictured bottom right. The author is pictured (coincidentally) at the centre.

After 2008, ISICs were moved to a 3-year cycle, but a hiatus due to reorganisation of the NATO Maritime capability group, and diminished interest from the UK (following withdrawal and cancellation of Nimrod MR2 and MRA4) and from other leading ASW nations, meant the 2014 conference was not held. With renewed interest in ASW in NATO, the conference has been expanded in scope and restarted from 2016 as the 'NUWIC' (NATO Underwater Warfare Interoperability Conference).

## COMPETITION IN SONOBUOY PROCUREMENT

Although UEL are now the sole UK supplier of sonobuoys, as the numbers of buoys being procured during the 80s started to rise the MOD introduced competition, inviting firms to incorporate their own PV designs and requiring firm price bids against a requirement (as opposed to building to a design specification).

At that time, sonobuoy work in the UK was primarily split between Plessey at Templecombe, and Dowty at Greenford. In the early 1980's MOD(PE) tried to keep both in business by splitting production between them. From the mid 1980's the companies were made to compete for all production and study work, which had the benefit of driving innovation.

Ultra Electronics Ltd. was taken over by the Dowty Group in 1977. It was acquired by the TI Group in 1992, and in 1993 became Ultra Electronics Ltd (UEL) again following a management buy-out led by Dr Julian Blogh. UEL became part of Ultra Electronics Holdings plc when it was floated on the London Stock Exchange in 1996.

The other main sonobuoy manufacturer was what was then the Plessey Company, subsequently taken over by GEC. Both UEL and Plessey were also invited to bid to manufacture the Barra sonobuoy (originally manufactured by Amalgamated Wireless of Australia). Competition was undoubtedly successful in reducing the price of buoys through the introduction of innovative new technology and manufacturing methods.

The RAE's work on procurement support for UK sonobuoys was carried out under a series of Sonobuoy Support Contracts with UK MOD which continued through DRA, DERA etc. The RAE also became the MOD's trusted expert on the uplink and, the receiver in the aircraft and for the airborne sonics.

By 1997 the UK company Ultra Electronics Holdings plc, had purchased US and Canadian sonobuoy companies, which became the Undersea Sensor Systems Inc (USA) and Maritime Systems Inc (Canada) businesses of Ultra, alongside Ultra's UK sonobuoy and sonar business, now known as Ultra Electronics Sonar Systems, and were in a position to supply sonobuoys to many nations.

Ultra Electronic Sonar Systems celebrated their half-millionth sonobuoy with a dinner at the RAC Club in Epsom. This event included on show a gold-plated G-size DIFAR (giving rise to a lot of jokes along the lines of "now we know why they are so expensive") and presentation of a first edition of "The Hunt for Red October", signed by the author Tom Clancy, to all the attendees. This book had caused quite a stir at the time because it contained many details of ASW which had up to that time been generally unknown.

In 1999 the author was present at UELs celebration event marking the delivery of their millionth sonobuoy, which included a Nimrod fly-past at their Greenford, West London site.

Celebrating the manufacture of

# 1 MILLION SONOBUOYS

by Sonar and Communication Systems

Over 50 years in
## SONOBUOYS

Ultra Brochure

By 2003 Ultra Electronics(UEL) had become the leading UK sonobuoy company and the UK MOD entered into a partnering agreement with them in November 2003 for the supply of sonobuoys and UEL became sole source suppliers to the MOD.

MOD/UEL Sonobuoy Partnering Ceremony- Chief of Defence Procurement Sir Robert Walmsley and Dr Julian Blogh
CEO Ultra Electronics sign at Abbey Wood November 2003      *UEL picture*

The main suppliers of airborne acoustic processing systems for the UK's airborne ASW assets were BAE Systems at Rochester (formerly a General Electric - Marconi Company) and Ultra Electronics Ltd (UEL). Prior to Nimrod's demise, UEL provided the processing for both the MR2 and the MRA4, in partnership with general Dynamics Canada (GDC), with BAE Systems supplying Merlin Mk1 and Nimrod MR1 airborne acoustic processing when it was part of GEC Marconi. Merlin Mk2 acoustic processing, which is supplied by Thales UK (TUK), is being upgraded to include multistatics in collaboration with Ultra. Prior to Nimrods demise, UEL provided the processing for both the MR2 and the MRA4 with BAE Systems supplying Merlin and previously earlier Nimrods when it was part of GEC Marconi.

QinetiQ (formerly RAE(F), DRA(F), DERA(F)) provided technical consultancy on behalf of the MOD for the GEC, UEL, and TUK systems, using their expert of long standing, Derrick McNeir, who gained high respect across the industry.

This general way operating continued into the 21st century until the demise of fixed wing ASW in the UK i.e. the retirement of the Nimrod MR2 MPA (March 2010) and the cancellation of its replacement the MRA4 (October 2010). After the cancellation QinetiQ's role continued, but now via the Merlin project, the only in-service UK air ASW asset in 2016.

# CHAPTER 9

## THE BASICS OF MODERN SONOBUOY DESIGN

This chapter was written in the main by Peter Evans ex-AUWE with some minor changes by myself.

It should be noted that the US developers of the first operational sonobuoy the omnidirectional AN/CRT-1 were unaware of most of these design considerations. In some ways the AN/CRT-1 could be described as an underwater microphone optimised for detection of acoustic frequencies available to the human ear.

Passive Sonobuoy Design Considerations:

Submarines emit noise from machinery (e.g. pumps, engines, propellers) and their movement through the water (turbulence, flow noise);

Submarine radiated noise needs to be detected against similar background noise, caused by the environment (shipping, breaking waves, wildlife) and sound generated by the receiving sonobuoy (electrical interference, flow and mechanical noise caused by movement of the sensor itself) etc;

Submarine detection relies upon being able to distinguish between these various noise sources, and the design of the sonobuoy can help to achieve this;

The designer seeks to minimise the sound caused by sonobuoy motion by careful decoupling of the hydrophone from the surface float. This is particularly challenging at low frequencies and requires an excellent understanding of the mechanics of noise generation, and the characteristics of materials used in the suspension system.

The designer can focus on acoustic frequencies where the differences between submarine and environ mental noise are the greatest. Submarine noise at low frequencies may, for example, be easier to detect at long ranges than higher frequencies. At short range, the converse may be true.

He can position his sensor at the optimum depth to enable detection of threat submarines. This will vary according to location and time of year, and the best depth can be determined through the use of sophisticated sonar modelling. As a result, sonobuoy operating depth needs to be adjustable within broad limits.

He can make his receiver directional, allowing him to reject noise from sources in other directions. This is analogous to cupping the ears to hear more clearly. There are a number of ways he can do this, but one of the best ways is to make the receiver physically larger, particularly in the horizontal plane. This is a major challenge for sonobuoy designers, as the large sensor needs to be packaged into a small canister to allow it to be deployed from an aircraft. Some designs achieve this through the use of folding arrays.

Achieving all of these requirements within a single design would require the development of extremely complicated and expensive sonobuoys. As a result, the designers have developed a range of buoys which are optimised to specific requirements and situations. The MPA will therefore carry a number of different sonobuoy types and will deploy these according to the needs of the mission.

Active Sonobuoy Design Considerations:

Active sonobuoys comprise of two components, the transmitter and the receiver. The transmitter generates high levels of sound in the water, and the receiver listens for echoes from objects in the vicinity. This is how dolphins and bats detect their prey.

Transmissions emitted by the active sonobuoy will reflect from every discontinuity in the water (the surface, waves, rocks, wrecks, seabed) as well as from the submarine itself. As a result, echoes from the submarine will need to be detected against this "reverberation". This is entirely analogous to shouting in an echo chamber, and trying to detect a single echo in the presence of multiple other echoes.

Echo detectability will vary according to frequency, duration and the type of transmission used by the sonobuoy. It will also depend upon the shape and construction of the submarine, and on the way it moves in the water.

In general, longer detection ranges are achieved using low frequencies and higher energy transmissions. These are stringent requirements for the sonobuoy designer, as low frequency transmitters are usually large in size, while high energy transmissions require significant power. This in turn makes huge demands on the battery.

The receiver component of the active sonobuoy has similar design requirements to passive sonobuoys:

The need for multiple operating depths, good decoupling from the sea surface and the use of large receiver arrays.

Packaging of both transmit and receive elements into a single canister is extremely difficult, and will inevitably lead to some compromises. As a result, the transmitter and receivers are sometimes packaged into separate buoys, allowing more flexibility in the design and potentially better performance.

The transmitter buoys tend to be much more expensive than the receivers, due to the high cost of large batteries and the powerful transducers required. With careful design, the receivers can be used both as passive sonobuoys and as the receivers for active buoys, thereby simplifying the sonobuoy inventory and providing cost economies.

# CHAPTER 10

## UK SONOBUOY DEVELOPMENT FROM THE 1970S TO THE 21ST CENTURY

## OVERVIEW

Apart from early Barra all these sonobuoys were developed by Dowty/Ultra Electronics and the specifications were written by the RAE etc. After the privatisation of QinetiQ, Ultra became the specification authority for new buoy types, including Barra SSQ981E and F, and ALFEA SSQ926).

- **1970s** - Initial Development of SSQ 963, Command Active Multi Beam Sonobuoy (CAMBS). First batch produced by Plessey.

- **1970s** - F-size Sonobuoy developed (SSQ 904 and SSQ 906) to meet Helicopter ASW needs.

- **1970s** - Initial Barra development in Australia. In 1975 Australia entered into an agreement with the United Kingdom relating to the Barra sonobuoy: Australian scientists design and manufacture this passive sonobuoy, while United Kingdom scientists design, develop and manufacture the airborne processor. 1977 AWA Limited in Australian awarded initial contract to produce Barra sonobuoys.

- **1970s** - G-size Directional Frequency Analysis and Recording (DIFAR) introduced.

- **1980** - First production Barra presented to the United Kingdom's High Commissioner for Australia. (25 February) Marks the beginning of deliveries of Barra sonobuoy to United Kingdom and Australian Air Forces and Navies

- The first UK Barra production contract (designated SSQ-981, based on the Australian SSQ-801 design) was awarded to Plessey Templecombe and the following SSQ-981A production contract was won by Dowty with a significantly simplified mechanical design, which deployed the array passively, with the arms opening gracefully, by gravity only, after the buoy had descended to depth.

- **1980/90s** - UK development of Barra, SSQ 981A, SSQ 981B

- **1980s** - Further Development of SSQ 963, Command Active Multi Beam Sonobuoy (CAMBS)

- **1990s** - SSQ 955, High Instantaneous Dynamic Range (HIDAR) Development, Replacement for DIFAR, Dual Mode (analogue and digital telemetry)

- **2003** – SSQ 963D CAMBS VI, Dual Mode (analogue and digital telemetry)

- **2003** – SSQ 906G Wideband LOFAR

- **2004** – SSQ 981E Barra, Dual Mode for passive and multistatic receiver

- **2004** – SSQ 955B, GPS added to HIDAR

- **2004** – SSQ 981F Barra-GPS Dual Mode. GPS added

- **2007** – SSQ 926, Active Low - Frequency Electro - Acoustic (ALFEA), Active buoy part of the UK's Multi-Static System

- **2007** - SSQ 937D, G-size Bathy Thermal Buoy

- **SB114** - Essentially a LOFAR buoy with an anchor attached and larger battery to extend the operating life. It was developed for Sweden for use in the fjords and has modifications to allow it to function in fresh water.

## THE CAMBS SONOBUOY

The UK's Mark 1C active sonobuoys were so large that they had to be carried in the aircraft's bomb bay. Air Staff Target 842 was issued in 1968 for a smaller A-sized active detection and localisation sonobuoy. Subsequent Studies at the Royal Aircraft Establishment, Farnborough in the 1970s showed that it was feasible to produce a high performance active sonobuoy in A-size. This resulted in the Command Active Multi-beam Sonobuoy (CAMBS) the SSQ 963, developed and produced by UEL to a specification produced by the RAE.

CAMBS has a novel cylindrical six stave volumetric receive array of a diameter which fits in an A-size sonobuoy and also does not deploy arms etc. on water entry and so is inherently reliable. The receive array was an RAE innovation and Ray Cyphus remembers that the original concept was the idea of Bill Camp a senior engineer at the RAE and he did the beamforming analysis. Derek McNeir says George Carter implemented the idea. I can remember him remarking that the problem with hydrophones is that they don't like getting wet! So the early array was contained in a cylindrical container that was filled with, I think, olive oil. He and Jack Hockey did the design and fabrication. Lloyd Jones coined the acronym CAMBS because he went to Cambridge University.'

Peter Viveash remembers. All the experimental work on the CAMBS was performed by RAE people Dr Lloyd Jones, George Carter, Ray Cyphus, me, and Jack Hockey. It is true that the CAMBS array was suspended in a can filled with oil

CAMBS entered service with the Nimrod MR2 in 1979 and is an active localisation, tracking and attack sonobuoy. Development of CAMBS has continued and the current version is the fully digital SSQ 963D, which remains in use with the RN's Merlin Mk2 aircraft. It is controlled by commands transmitted from the aircraft and has four selectable operating depths.

The UEL picture below shows the CAMBS transmit line array and its six stave volumetric receive array.

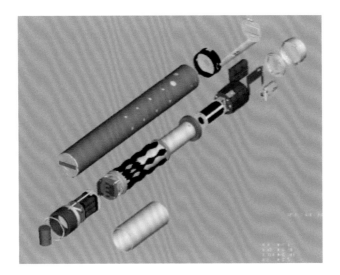

CAMBS    *Ultra pictures*

CAMBS 963D enhancements over previous versions

- Increased source level / effective search area to counter clad SSK targets using countermeasures.

- Dual-Mode RF telemetry - digital and analogue (for backwards compatibility).

- Digital telemetry - high dynamic range and adaptive beamforming in the MPA or helicopter would enable more effective Acoustic Counter-Counter Measure (ACCM) capability.

- More powerful pings (+6dB) enabled by pushing current transducer, amplifier & battery technology for improved detection on first ping.

- Pulse shaping for reduced reverberation side lobes.

- More ping types and dual acoustic frequencies, enabling more buoys to ping at same time (multi-monostatic operation).

- 1hr 'Attack' Mode at high source level.

- 4hr life 'training' Mode at lower source level to reduce expenditure.

| Design Improvements | Operational Benefits |
|---|---|
| Digital telemetry | Dynamic range enhanced from 40dB to 80dB<br>Allows the processor to counter acoustic countermeasures<br>Growth for future sensor data and buoy status telemetry (e.g. depth) |
| Digitally generated shaped acoustic pulses | Improved detection performance:<br>▪ Against slow targets by control of the reverberation roll-off<br>▪ Through improved control of pulse shape and therefore pulse matching |
| Multiple acoustic frequencies | Two buoys can use different acoustic frequencies without mutual interference |

# THE BARRA SONOBUOY

Also in 1979 the SSQ 981, "Barra," an Amalgamated Wireless of Australasia (AWA) designed A-size directional passive sonobuoy was accepted into service. The first UK Barra production contract (designated SSQ-981, based on the Australian SSQ-801 design) was awarded to Plessey Templecombe, but the following SSQ-981A production contract was won by Dowty with a significantly simplified mechanical design, which deployed the array passively, with the arms opening gracefully, by gravity only, after the buoy had descended to depth.

Barra has a 5 arm planar receive array for passive detection and was redesigned in the UK in 1990 by UEL to become the SSQ 981A. The current version is the UEL produced dual Mode SSQ981E which can operate as a receiver in an active multistatic sonobuoy field or as a passive receive array and has adaptive digital signal processing. The advantages of a digital architecture are: better performance in high shipping and rain noise and multistatic environments and an improved radio range.

The pictures below courtesy of UEL.

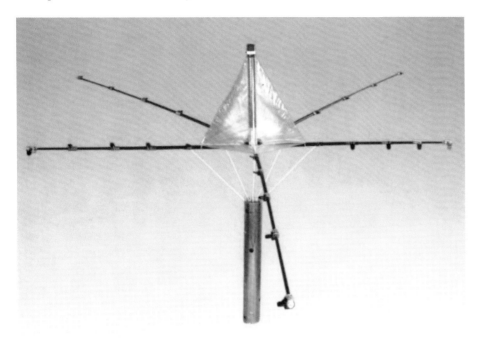

Barra SSQ 981E deployed    *UEL Picture*

Barra SSQ 981E deployed in tank    *UEL Picture*

SSQ981E undeployed    *UEL Picture*

# DIFAR AND HIDAR SONOBUOYS

The first British **Directional Frequency Analysis and Recording (DIFAR)** sonobuoy was produced in 1985 by UEL. This was constructed in G-size, (419.10 mm long X 123.825mm diameter) and combined LOFAR techniques with a directional sensor. It replaced the omni-directional SSQ 904 sonobuoy, and became the principal passive sonobuoy in use for UK airborne ASW operations in the 1990s.

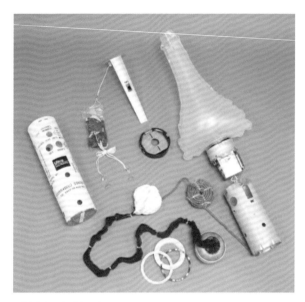

SSQ954D DIFAR

The picture above illustrates the components of a SSQ954D DIFAR which are as follows:

- DIFAR sensor,
- Short suspension,
- Kite/damper (packed),
- Main compliant suspension,
- Cable pack,
- Surface unit (for flotation, further signal processing, and telemetry),
- Deployment plate (to retain the components until water entry),
- Parachute,
- Wind-flap (which deploys the parachute),
- Internal packing rings and springs.

The pictures and text courtesy of UEL.

The current G-size UK UEL DIFAR design is **HIDAR, (High Instantaneous Dynamic Range)** which has all digital electronics. Its high dynamic range and linearity allows passive operations in high ambient noise conditions and as a low frequency active receiver. The VHF telemetry can be selected between digital and standard FM analogue formats to ensure interoperability with other forces.

Fairly soon after ISIC 2002, in 2003, the MOD updated its MPA Airborne ASW strategy and all work on improvements to passive buoys and CAMBS was cancelled. What survived, however, was very important and that was the UK Multistatics Programme.

HIDAR *UEL Picture*

The picture below shows the current range of UEL sonobuoys:

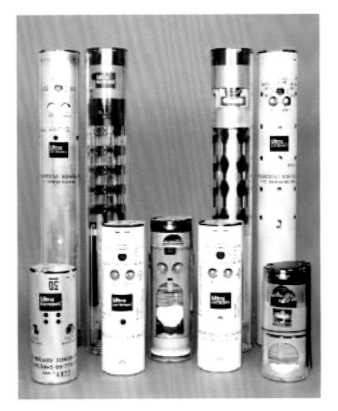

*UEL Picture*

# CHAPTER 11

## THE THIRTY YEAR DEVELOPMENT OF A UK MULTISTATICS SYSTEM (SEARCH FIELDS OF SEPARATE ACTIVE TRANSMIT AND RECEIVE SONOBUOYS)

## THE NEED FOR MULTI STATISTICS

During the Cold War, the early Soviet submarines were relatively easy to detect. However, they became progressively quieter, especially after 1968, once the Walker spy ring revealed the existence of SOSUS.

**SOSUS** is an acronym for **sound surveillance system** and is a chain of underwater listening posts located around the world in places such as the Atlantic Ocean near Greenland, Iceland and the United Kingdom—the GIUK gap, and at various locations in the Pacific Ocean. The United States Navy's intent for the system was for tracking Soviet submarines.

As a result of the Walker spy ring's activities, by 1979 the USSR had fielded the Victor 3 which was significantly quieter than previous Soviet nuclears and was more difficult to detect passively. The Akula followed from 1984 and was quieter than all previous Soviet designs.

Victor 3 submarine    *Author's Collection*

Akula submarine     *Author's Collection*

The shape of new design Soviet submarines also changed and sound absorbing tiles were added to their hulls. These changes were mainly aimed at reducing their monostatic target strength particularly at torpedo sonar and fire control sonar frequencies. These changes are analogous to stealth in military aircraft.

Once the Cold War ended, the focus of UK ASW shifted to very quiet SSKs operated by nations which could be a future threat. The development of very quiet diesel (SSK) submarines made passive detection progressively less effective, despite increasing the number of buoys deployed, with 32 channel and even 64 channel airborne processors becoming common with fields of up 64 passive sonobuoys.

Iranian Midget Submarines - *UEL picture*

With the proliferation of small quiet submarines, interest grew in the US and UK in the 1980s in a new way of using airborne sonar as a wide area search sensor, using multistatic sonobuoy fields employing transmit sonobuoys and separate receive sonobuoys. In the UK, this led to research associated with the development of a multistatic system called the Active Search Sonobuoy System (ASSS), also known as SR(SA)903 (Staff Requirement Sea/Air 903).

At the date of writing, current Russian SSNs have started patrolling again near UK waters according to media reports, and continue to represent a difficult problem for passive sonobuoy detection.

The ASSS concept of employment involved an MPA or helicopter sowing a field of active transmit sonobuoys and separate receive buoys, either HIDAR or Barra. The aircraft would contain appropriate signal processing and displays to process the acoustic information received from the HIDAR or Barra.

Picture illustrating Multistatic Sonobuoy Search *courtesy of UEL* - not to scale

A UEL presentation available on the internet discusses the potential advantages to Multistatic Sonobuoy Search, compared to traditional passive or monostatic sonobuoy operations:

## MULTISTATIC SONOBUOY SEARCH ADVANTAGES

Probability of detection. When the transmitter and receiver are co-located, the energy from the transmitter must travel to and from the target along the same path. However, the strongest return is generally not back towards the source, and so separating the receiver and transmitter can increase the probability of detecting the target. In any case, monostatic and multistatic, active detection ranges will usually far exceed passive detection ranges on quiet submarines.

**Search/Coverage area.** Using a field of widely separated transmitters and receivers provides many detection opportunities, and so provides detection and tracking of targets over a much greater search/coverage area than could be obtained using a monostatic system. The use of multistatics can reduce the number of active sonobuoys required to search an area. These buoys are typically much more expensive than the receivers, so the overall cost of a mission is thereby reduced.

**Tactical advantage.** Operationally, the separation of the transmitter from the receiver is unknown to the target submarine. A submarine would traditionally assume that the detection threat is in the direction of the sound source, whereas the receiver may be in an entirely different direction, and potentially much closer to the submarine than may be imagined. The submarine's lack of knowledge of the receiver positions complicates its ability to evade detection through appropriate manoeuvres

Also the transmit and receive buoys may be at various depths and relative ranges, making it very difficult for the submarine to exploit acoustic propagation conditions to avoid detection.

## UK MULTISTATICS

In 1988, Peter Evans took over as section leader at AUWE Portland, working with Willie Wildash, Simon Vines and Geoff Searing on ADS (Active Dipping Sonar) and sonobuoy research. Sonobuoy emphasis had now switched to active, and the work at Portland was now focused on assessing the applicability and likely performance of bistatics for sonobuoy use. A large modular passive receive array was modified for use as a bistatic receiver, and a modular vertical array sound source was developed for trials use. The two arrays were free-floating and were deployed from a ship, which then stood off some 5 miles away. Measurements were then carried out until either the batteries ran out, or until the arrays drifted into an unsuitable configuration. A bistatic sonar model was developed by Geoff Searing in collaboration with Plessey Templecombe, and model characteristics and assumptions were validated against the trials data. As part of the research, low frequency "bender" transducers were developed by Dowty, and this technology was subsequently adopted in other higher frequency systems.

## SR (SA) 903 FEASIBILITY STUDY – ACTIVE SEARCH SONOBUOY SYSTEM (ASSS)

In the late 1980s, the RAE carried out an ASSS Feasibility Study in support of Staff Requirement SR(SA)903. The Feasibility Study was conducted by a RAE team led by Ray Cyphus, RAE's active sonobuoy expert, with support from AUWE Portland and UK industry, primarily Plessey Templecombe.

The RAE placed a contract with Plessey Templecombe which focused on the use of large numbers of active sonobuoys operated primarily as a field of monostatic sensors. A large number of sensors were to be deployed in each field, and the sources were "pinged" in a random or programmed fashion, increasing the likelihood of detecting the broadside "glint" from a transiting submarine. Plessey studied the design requirements and operating philosophy of the overall system, and developed models to predict performance and area search effectiveness.

A typical answer the area search model produced was:

For a sonobuoy field area search of maximum duration of x hours using y candidate transmit and z receive sonobuoys and airborne processing, the probability of the field detecting submarine type AA is say 0.7 for a specified sea area size and sonar environmental conditions.
The model was used to assess the area search performance of the then candidate ASSS systems to meet SR (SA) 903.

The solution recommended from the Feasibility Study required a new five arm receive sonobuoy and a new transmit sonobuoy operating in the 2-4 KHz region. Acoustic processing options in detail were not covered in the study although it was acknowledged that novel geographic based techniques would be required for multistatic sonobuoy field displays.

This work did not lead on to a Project Definition Study at that time as there were a number of risk areas identified and a lack of supporting at sea performance data.

Also the predicted search field performance was rather disappointing and in retrospect the first 903 FS was carried out at a time when the technology available was not available to successfully search tactically useful sized areas. It must be emphasised that this comment is made with the benefit of hindsight.

The 903 Feasibility Study work carried out by Plessey looked at the use of large active sonobuoy fields for the detection of submarines, and in response to this, research at Portland switched to the use of multistatics as a possible detection method. Work now focused on the use of Barra-style arrays for multistatic detection, and the use of smaller, higher-frequency bender arrays as the source. At that time, Barra arrays were hard-clipped (each hydrophone transmitted a 1 or a 0 representing positive or negative acoustic pressure, with no amplitude information), so Willie Wildash instigated research to investigate the use of multi-bit sampling and processing within the existing Barra array. The existing hydrophones and processing were replaced with multi-bit receivers, and the data was then streamed back to the ship using the existing sonobuoy transmission standards.

The system operated in one of two modes. In its default state, the system operated in an identical way to Barra, returning clipped data over the whole passive operating band. On detection of a large active transmission, the system switched to multi-bit operation, returning high resolution data for a predetermined time. Processing was carried out over a narrow frequency band, centred on the operating frequency of the source. In this way, total bit rate could be constrained within existing Barra operating frequencies. The additional processing was implemented within the existing array design, although the array itself had to be extended and deployed from a ship, rather than being air-launched. A number of successful trials were carried out using the system.

In parallel with the experimental work, a multistatic system performance model was also developed to investigate the likely performance benefits of multistatics. This made use of performance predictions from the existing bistatic sonar model. A number of other studies were carried out, looking at the feasibility of using other sources, such as ADS, LFAS (a Low Frequency Active Surface Ship deployed towed source) and existing hull mounted surface ship sonars, to activate multistatic receivers.

Other AUWE work focused on bistatics, developing an understanding of how reverberation and detection performance varied with source, receiver and target position. This required the development of a novel bistatic reverberation model and the conduct of a number of sea trials involving submarines. The types of source and receiver used in the two research programmes were similar, comprising a vertical array of projectors for the source, and a large horizontal planar array as a receiver. The initial AUWE work focused on relatively low frequencies, and included the development of Bender transducers with the appropriate power. The use of low frequencies required a correspondingly large receiver array, which was too large to be deployed from an aircraft.

The RAE system, on the other hand, was intended from the outset for air deployment, so used higher frequencies and smaller arrays. The two programmes gradually converged, with efforts being focused on the use of higher frequencies and smaller arrays, and the development of the "multistatic" concept (vs. monostatic or bistatic), where each transmission was exploited by all the receivers in the field, rather than by a single receiver only. As part of this activity, the AUWE programme looked at the feasibility of modifying existing Barra sonobuoys to act as large aperture multistatic receivers.

## POST SR (SA) 903 FEASIBILITY STUDY RESEARCH

A number of derisking research activities were started by the RAE and the AUWE after the SR (SA) 903 Feasibility Study, and included producing and trialling prototype transmit and receive sonobuoys and obtaining at sea performance data.

The RAE contracted UK industry to produce an Experimental Active transmit and receive Sonobuoy (ESE). ESE was built to be deployed from a surface ship for at sea trials and contained a 5 arm receive array and an active transmitter operating in the 2 to 4 KHz region. A series of trials were conducted by the RAE between 1989 and 1995 employing ESE and much useful data was gathered.

In the picture ESE is being deployed from DERA's Colonel Templer research vessel during a trial. The five arms are the receive array with 8 hydrophones per arm. The transmit array is contained in the cylindrical section.

RAE ESE Experimental Sonobuoy – *Authors collection*

AUWE/DRA Portland also carried out supporting transmit and receive sonobuoy research as part of the SR (SA) 903 derisking activity. An experimental coherent active transmit vertical line array sonobuoy was designed and built at Portland with Ultra providing the amplifiers and individual transmitters known as benders. This was known as ESS (Experimental Sound Source), and transmitted at 1300Hz. An echo repeater, BERT (Bistatic Echo Repeater Transponder), was also produced in house. This equipment was used in a series of bistatic trials from 1986 onwards. Ultra further developed the bender transmitters and they are utilised in the Ultra's current ALFEA active transmit sonobuoy, as well as evolving for use in horizontal towed active sonar arrays.

Portland also designed and built VERA, Versatile Experimental Receive Array. VERAs purpose was to investigate receive array performance as a multistatic receiver buoy at different transmit buoy frequencies using different numbers of receive hydrophones and geometries.

The DRA at Farnborough investigated acoustic processing/display techniques known as energy maps employing geographic displays based on an approach proposed by DRAs airborne processing guru Derek McNeir. This was a first step away from receive sonobuoy centred displays and Plessey's Roke Manor Research unit carried out this work.

The bistatic target strength assumptions adopted in the 903 Feasibility Study were first pass estimates and because of their importance extensive work was then started by AUWE/DRA Portland on their research programme to better define the bistatic target strength of potential target submarines.

As part of the DRA's long term research programme at Farnborough the development of a more detailed computer model than the one used in the 903 Feasibility Study was begun, able to assess the probability of a multistatic field of sonobuoys detecting submarines with more confidence. This model was known as ASSAM and after further development was eventually used by the MOD and industry to support the ASSS programme.

## ASSS PROGRAMME RESTARTS IN 1995 AS THE SR(SA)903 REVALIDATION STUDY

A relook at ASSS in 1995 was initiated by MOD and was stimulated by information received in DOR(Air) about the SSQ 110, a US impulsive sound source (which used small explosive charges). The Feasibility Study associated with SR(SA) 903 would not have been re-opened otherwise. US studies on the area search performance of SSQ110/DIFAR multistatic search fields appeared quite promising at that time. The USA was ahead of the UK on certain aspects of multistatics in the mid 1990s and had developed the SSQ110 an impulsive multistatic transmit buoy for use with DIFAR as a multistatic receiver. It was undergoing extensive trials in the US using P3s to deploy multistatic SSQ110/DIFAR search fields.

The ASSS programme restarted with DRA Farnborough leading with the support of UEL and

GEC. The USA supported the programme under a data exchange agreement with DRA Farnborough on behalf of UK MOD.

The author was the DRA project manager and was supported by Mike Clapp, formerly of Westland System Assessment Limited (WSAL), and DRA colleagues at both Farnborough and AUWE.

An important early tasked carried out by the author and Mike Clapp was to redraft for DOR(Air) the SR (SA) 903 performance requirements in terms of area search effectiveness rather than detection range.

DRA purchased some SSQ-110s on behalf of UK MOD and four evaluation trials took between July 1995 and May 1998 using Nimrod MR2s and DIFAR and Barra as the receive sonobuoys. These trials were analysed by Dr Ben Wynne's processing team at DERA Farnborough. A number of different airborne processing systems were used in these trials and in the subsequent analysis. These included systems provided by BBN, DSR, UEL, GEC and DRA/DERA. When the Revalidation study took place only the BBN results were available. As part of the APR (Applied Research Programme), QinetiQ at Winfrith, led by Neil Skelland and Simon Vines, analysed some of the impulsive trials data using an analysis system developed previously on the ARP. This system was aimed at distinguishing between sub and non-sub returns as a potential acoustic operator decision aid. The results were quite promising

The Revalidation Study was essentially a qualitative and quantitative review of multistatic sonobuoy systems that could meet the requirements of SR(SA)903 with low risk.

A classic, in OR terms, cost effectiveness assessment was carried out as was some ground breaking Environmental Impact Assessments. The main tasks were:

1. Review the original SR (SA) 903 Feasibility Study and more recent research activities and identify current or near future sonobuoy and multistatic processing and display developments which could form part of a low risk ASSS. Include US developments, including both SSQ110 impulsive and electroacoustic transmit sonobuoys and the ADAR receive sonobuoy.
2. Review the predicted sonar conditions in the geographical operating areas specified in SR (SA) 903 in terms of their likely influence on search effectiveness.
3. Carry out Environmental Impact Assessments for sea areas where multistatic trials are to take place employing SSQ110s. DRA Winfrith carried out his task.
4. Develop a computer based simulation model able to predict the probability of a multistatic field of sonobuoys detecting submarines for both impulsive and electroacoustic transmitters. Plessey Templecombe were contracted to modify the model developed in the 903 Feasibility Study for this task.
5. Select low risk candidate multistatic sonobuoy search systems and assess their area search performance in the scenarios specified in SR (SA) 903 using a suitable computer model. Both impulsive and electroacoustic systems are to be included.

6. Review and learn lessons from SSQ110 multistatic trials analysed by BBN using their multistatic processing system.
7. Carry out a risk assessment of the candidate systems.
8. Using the outcome of activities 1 to 7 give advice to MOD as to whether SR(SA) 903 merits proceeding to a PD phase

Task 3 led to the development by Dr Tony Heathershaw and an expert multidisciplinary team at DERA Winfrith and later at QinetiQ's Environmental Impact Assessment Centre, at the Southampton Oceanographic Centre, of ground breaking techniques for assessing the impact of sound on sea mammals and fish that takes into account sound frequency, intensity and duration. The technique was based on health and safety regulations for humans. In a newspaper article Dr Heathershaw was quoted as saying. *The lower the noise, the longer a person can continue working; the higher and louder the noise, the less time a person should spend near it.* "We have every reason to believe that work on human beings can be applied to marine mammals as well," "We may not know what's in the ocean at any one time, and we may not know what the threshold for hearing in a particular species in the ocean at that time is. The approach we take is very precautionary."

UK Industry has adopted Dr Heathershaw's mitigation measures, and Environmental Risk Assessments and mitigation remains a very important consideration in the design and operation of such systems.

The outcome of activity 8 was that SR(SA)903 merits proceeding to a PD phase, but that significant risks remained in the multistatic processing and target classification area and that safety and environmental impact issues with impulsive sources might preclude their carriage in UK MPAs. It was also suggested that meeting the SR (SA) 903 search coverage requirements in all the sonar environments as specified in 1995 would be challenging.

The outcome of task 5 was that in terms of area search performance the following options were considered worthy of further investigation: UK electroacoustic sonobuoy options similar to UELs current ALFEA, US SSQ110 impulsive based options, and US electroacoustic active sonobuoy (now known as the SSQ125) based options.

A number of the sonobuoy options were at the prototype stage i.e. were immature technologies in the mid to late 1990s.

The US options included ADAR then a new and unproven complex receive sonobuoy with 40 hydrophones, and the US ADLFP (Advanced Development Low Frequency Projector) electroacoustic active sonobuoy, now known as the SSQ125, which was at an early stage of development.

During 1997/8 at DERA Farnborough Ben Wynnes' processing team developed the world's first impulsive energy map system using a software package called PVWAVE and used data from the UK's SSQ110 evaluation trials to test and demonstrate their solution.

After the 903 Revalidation study was complete and the findings were reported, the UK MOD decided to proceed with a 903 ASSS PD phase to meet URD2006. DERA wrote the technical requirement for the MOD's ASSS PD study Request for Quotations and derived the technical assessment marking scheme. A team of DERA experts from both Farnborough and Winfrith carried out the technical assessment and then assisted the MOD in managing the technical aspects of the PD study

Ultra and TMSL conducted parallel 18-month ASSS Project Definition Studies, which were completed in November/December 1998.

After the PD studies had reported, the ASSS programme entered a Risk Reduction Phase in 1999/2000 because the MoD perceived that the understanding of multistatic performance was not good enough and the risks were therefore too high to proceed. DERA's contribution to the Risk Reduction Phase was as follows:

- Derivation of Bistatic Target Echo Strength data through trials and modelling.
- Organising data gathering trials using both DERA and consortia equipment.
- Conducting Environmental Impact Assessments.
- Development of the ASSAM model to support the BAE Systems bid for the ASSS Demonstration and Manufacture Phase that would lead to the ASSS being integrated in the Nimrod MRA4. This was done under the direction of an Integrated Modelling Group, chaired by BAE Systems and with input from the consortia and the MOD.
- Support to COEIA (Combined Operational Effectiveness and Investment Appraisal) activities.

## A PROTOTYPE APPLICATION OF MULTISTATIC ENERGY MAPS TO THE CAMBS ACTIVE LOCALISATION SONOBUOY

During the SR(SA) 903 Risk Reduction Phase, DERA developed a PC based Multistatic CAMBS Energy Map Demonstrator as a low cost way of investigating the benefits of energy maps by applying the technique to the CAMBS active localisation sonobuoy.

The demonstrator was deployed during a trial with a Nimrod deploying three CAMBs buoys and a UK SSN as shown below.

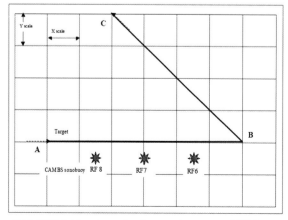

CAMBS Multistatic mode of operation          *UEL image*

The demonstrator mode of operation derived was as follows:

- One CAMBS acoustic frequency was used for all sonobuoys.
- Ping sequencing was employed to avoid mutual interference between transmitting sonobuoys.
- Buoy channels not pinging apply 'Active' processing.
- Increased data rate archived compared with monostatic operation due to multiple processing from each ping.
- Detection performance per ping not limited to monostatic, but available on all buoys.

An example CAMBS CW PING energy map is shown below. Energy levels are denoted by red to yellow as energy increases. Yellow usually denotes a contact. The contact in this case is a submarine but energy maps can also detect bottom features. However, these are static and can be recognised for that reason.

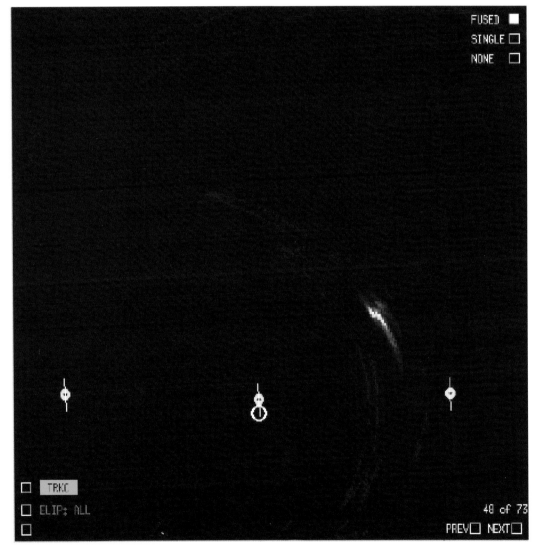

CW Ping energy map     *UEL image*

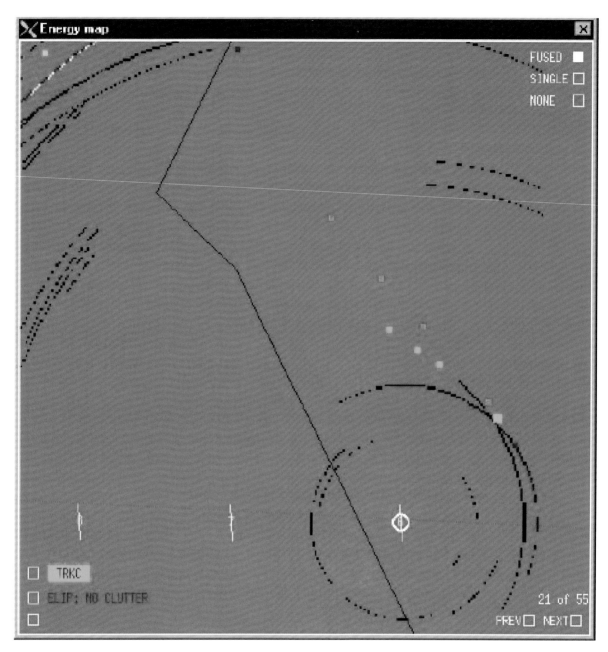

CAMBS FM ENERGY MAP    *UEL image*

When multistatic FM contacts are available from more than buoy then the high resolution FM Mode allows targets to be pinpointed. The distant ellipses are bottom bounces. The green /purple squares are contacts.

Trial detection performance outcome:

- FM - improved significantly in poor detection conditions - sensitive to buoy pattern and target broadside angle.
- CW – small but worthwhile performance improvement even in favourable detection conditions for monostatic (specular glint).
- FM Mode - use of specular returns, target held for longer.
- Intersection of ellipses', particularly in FM Mode can provide accurate localisation.
- Situational awareness can be improved through geographic mapping

## UK MULTISTATIC ACTIVITIES FROM 1999 TO THE PRESENT DAY

During the Risk Reduction Phase, DERA deployed Mike Clapp to BAE SYSTEMS Warton from October 1999 to October 2000 to provide input to the production of the ASSS User Requirements Document (URD) and the Invitation to Tender (ITT) and associated specification. SR (A) 903 formally became URD 2006 in September 2000.

During this period TMSL decided to pull out of the ASSS programme, one of the reasons reportedly being the difficulty in accessing system interface information.
The 2-year Risk Reduction programme, culminated in the specification and proposal of a system upgrade for MRA4 by Ultra and BAE SYSTEMS to implement ASSS

The MOD then decided in November 2001 not to proceed with ASSS as proposed by BAE SYSTEMS and the ASSS Main Gate submission, planned for 2001, was placed on hold for 2 years by MOD pending a review of affordability and to allow additional technologies to be evaluated. Also impulsive systems were dropped as they were assessed to be too difficult from a logistic, safety and infra-structure perspective and posed significant environmental impact issues. Also high false alarms rates were seen as a major risk area.

The procurement of ASSS, as conceived at the time, was stopped because of the perceived high cost of integration in the Nimrod MRA4 and its acceptance into service. Another programme was pursued instead that had the potential of being more cost effective. This alternative programme proposed an area search component and a cued search/localisation component. The cued search component was an electro-acoustic system broadly similar to ASSS and now referred to as the Multi-Static Active (MSA) system. It employed UEL's ALFEA as the active transmit buoy and HIDAR or Barra as receivers.

The main contracts conducted by industry related to SR(SA)903 from 1997, also known as 'ASSS' and 'MSA', are listed below.

| Date | Project Title | Customer | Prime | Sub-contractors |
|------|--------------|----------|-------|-----------------|
| 1997-1999 | Project Definition (PD) study | MOD | TMSL | Not known |
| 1997-1999 | Project Definition (PD) study | MOD | Ultra | Roke Manor Research, Westland Systems Assessment, Lockheed-Martin Sanders, ERAPSCO, DERA |
| 1999-2000 | Risk Reduction (RR) study | BAE SYSTEMS | Ultra | CDC, Engage (formerly WSA), DERA |
| 2000 | Full development proposal submission | BAE SYSTEMS | Ultra | GDC (formerly CDC) |
| 2003-2006 | MSA Engineering / Technology Demonstration (ETD) | MOD | Ultra | GDC |
| 2007-2009 | Extended Assessment (EA)/ Capability Demonstration | MOD | Ultra | GDC |
| 2010-2010 | Further Assessment (FA) | BAE SYSTEMS | Ultra | GDC |

QinetiQ, which took over most of DERA's activities in 2001, provided support to the Nimrod IPT during the Ultra MSA Engineering / Technology Demonstration (ETD) contract.

A COEIA associated with a large area search system was conducted by Dstl in 2004/2005 and showed that MSA could provide a cost effective localisation option for prosecuting potential targets detected by the alternative technology large area search component. Mike Clapp from QinetiQ managed the MSA part of the COEIA that used the ASSAM model previously developed to support the ASSS programme.

The development of MSA continued from 2007 to 2009 as an Extended Assessment/Capability Demonstration Phase in which QinetiQ organised a series of trials of MSA deployed from a Nimrod MR2.

The contracts let to Ultra from 2003 to 2010 included using the Nimrod MR2 as a risk reduction platform, resulting in it having an initial operational multistatic search capability at the time it left service in 2010

It was intended to include a variant of MSA in the Nimrod MRA4, produced by Ultra Electronics, with GD Canada providing the processing, but the aircraft programme was cancelled in 2010.

Since the demise of the Nimrod MRA4, work continued on MSA through the Merlin prime contractor Lockheed Martin, supported by Thales UK and Ultra Electronics, with intention of it being installed in the Merlin Mk2 helicopter.

A similar multistatics programme has been conducted in the US but for MPAs, and is now known as MAC (Multistatic Active Coherent) and uses the SSQ-125 sonobuoy – a joint venture of Ultra and Sparton with the Active Receiver (ADAR) sonobuoy.

The contracts let to Ultra from 2003 included using Nimrod MR2 as a risk reduction platform, resulting in it having an initial operational multistatic search capability at the time it left service in 2010.

Two final operational exercise flights of the Nimrod MR2 took place on the 24th and 25th March 2010. Both flights choose to use the MSA system, with ALFEA and HIDAR buoys, and MSA software installed on the UEL/GDC ASQ971 acoustic processor. Both flights successfully detected and tracked a friendly (UK) submarine target using MSA, and then launched successful active attacks using CAMBS. The picture below shows a screenshot of the buoy map and target track from the penultimate exercise, showing 27 HIDAR buoys ('H'), 5 ALFEA buoys ('F') and the target track, prior to the active attack. The last in-service flight of Nimrod MR2 was a flypast by two aircraft at RAF Kinloss on the 26th March 2010. Nimrod MR2 was formally withdrawn from service on 31st March 2010.

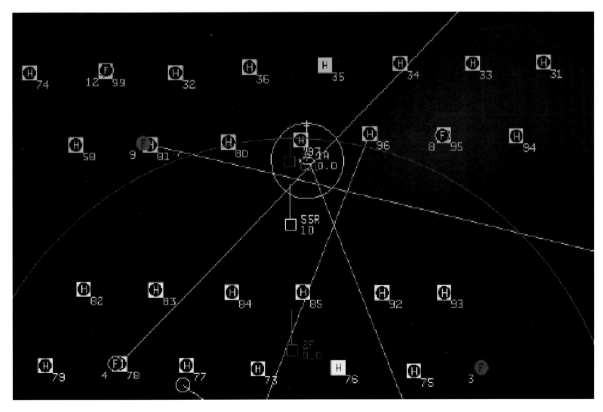

*UEL image*

The main features of the ALFEA component of Ultra's MSA system is as follows:

# ALFEA - ACTIVE LOW FREQUENCY ELECTRO-ACOUSTIC SONOBUOY SSQ 926

Features

A-Size, Variable Power, 1-2 kHz Source, Programmable waveform types, Autonomous Ping capability, Robust to RF interference, 4 depth capability, Autonomous/Electronic Function Select, Command Function Select, GPS fitted

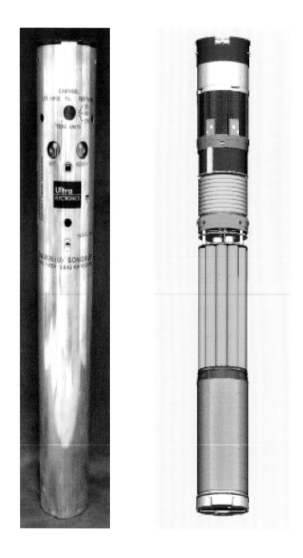

SSQ 926 ALFEA - *UEL PICTUREs*

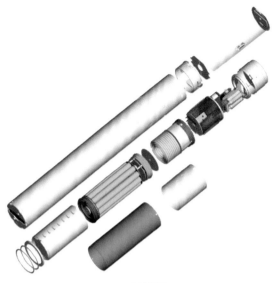

SSQ 926 ALFEA - *UEL PICTUREs*

# WHY DID THE UK MULTISTATICS PROGRAMME TAKE SO LONG?

Part of the reason is that the first SR(SA)903 Feasibility Study was carried out in the 1980s at a time when the technology available was not available to search areas of a tactically useful size. It must **be emphasised that this a comment made with the benefit of hindsight.**

There was then a long delay before the Feasibility Study was revisited during which time further research was being conducted in the UK and independently in the US. It was the US investigation and development of impulsive sources which appeared to be promising performance wise which was the main driver for the new look at multistatics in 1995. Six years later Ultra and BAE SYSTEMS submitted a proposal to MOD to implement ASSS on the MRA4 as a system upgrade.

The MOD then decided in November 2001 not to proceed with ASSS as proposed by BAE SYSTEMS and the ASSS Main Gate submission, planned for 2001, was placed on hold for 2 years by MOD pending a review of affordability and to allow additional technologies to be evaluated. At the same time a number of issues led to impulsive sources being ruled out in 2001:

- False alarm rates were high compared with electroacoustic options and difficult to manage; the use of a small number of impulsive pings is not conducive to persistent surveillance or tracking, which is a strong false alarm reduction aid

- Environmental impact and safety issues were a bigger cause for concern compared with electroacoustic sources.

The next significant delay was caused by the MOD's decision in November 2001 not to proceed with ASSS for cost and risk reasons. However, there was a resolve in the MOD to benefit from the investment that had been made in ASSS and this led to the MSA ETD in 2003 and the subsequent activities already described.

A cut down version of ASSS known as MSA was then developed with the aim of becoming a cued search/localisation system for the MRA4. MSA was an electro-acoustic system broadly similar to ASSS and referred to as the Multi-Static Active (MSA) system. It employed UEL's ALFEA as the active transmit buoy and HIDAR or Barra as receivers.

MSA was evaluated on the Nimrod MR2 and was deployed against a real submarine during the ETD phase, on 3rd February 2005, successfully detecting a friendly diesel (SSK) submarine. Further successful trials series took place later that year, in 2006, 2008 and 2010. However, by the time electroacoustic multistatic systems (now known as MSA) were first being operationally trialled, an ULTRA system to which proved the concept more than 10 years ago, the MOD had downgraded Airborne ASW in its list of priorities. This system was developed by Ultra and is known as MSA. Since then progress on MSA has been very slow.

A major factor in the faltering progress of MSA, via a seemingly interminable and contrived series of R&D contracts, was that it coincided with the troubled Nimrod MRA4 development programme. Nimrod MR2 benefited from MSA, but officially only as a research or 'de-risking' programme, as no new capability could be added to an aircraft in the process of being withdrawn. Meanwhile, the risk of impacting the already delayed main Nimrod MRA4 programme made the introduction of a significant capability upgrade mid-contract programmatically too difficult.

The result of this has been a very long and drawn out programme and only now has a cut down version of the planned fixed-wing version of the UK's multistatic system, MSA, been implemented on the Merlin helicopter.

The cancellation of the MRA4 and the retirement of the MR2 in 2010 appeared at the time to put an end to fixed wing multistatics but the advent of the P8, and future fixed wing UAVs, will change all that.

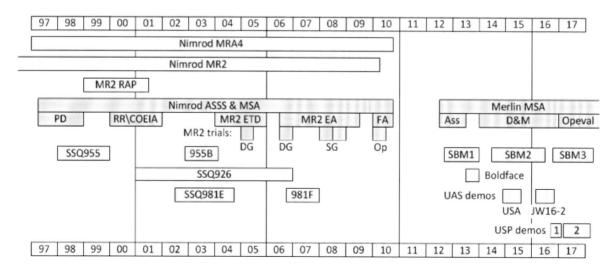

ASSS Programme from 1997 *from UEL*

# CHAPTER 12

## DERA IS SPLIT UP IN 2001 AND PRIVATISED QINETIQ AND MOD AGENCY DSTL ARE BORN

In 2001, the majority of DERA became QinetiQ, a private company and the minority became Dstl and remained as a MOD government agency. Three senior DERA Airborne Maritime Systems staff at Farnborough became members of Dstl and the rest became part of QinetiQ.

Dstl carried on the traditional role of RAE/DRA/DERA as MOD's trusted internal technical advisor, but with much reduced staffing such that a lot of research was contracted to private industry including QinetiQ.

QinetiQ retained some of its trusted advisor roles after privatisation, but as would be expected the relationship with MOD and Dstl gradually became more distant and a lot of its previous DERA work was opened up to competition.

By 2001, MOD's Sonobuoy Systems Research Programme had been drastically reduced, as by then it was 10 years since the demise of the USSR (the USSR was dissolved in December 1991) and the submarine threat to the UK was perceived as being much reduced.

This was reflected in the 2001 capability statement of MASP, which was part of QinetiQ at Farnborough's new Sensor and Avionic Systems Department (SASD). SASD was QinetiQ's centre of excellence for avionics and airborne sensors systems.

## MASP CAPABILITY STATEMENT

Based on my memory I believe there were two capability areas in 2001 with the staff based in the main at Farnborough:

- Sonobuoy Sensor Systems
- Maritime Airborne Processing

The majority of people in these capability areas comprised QinetiQ staff, but there were a few self-employed senior consultants to fill skill gaps. The staff were deployed on a multitude of projects each managed by an APM (Association for Project Management) qualified project manager, who was responsible for meeting the customers' contract requirements on time, budget, quality standards etc and where possible exceeding his expectations. Where needed QinetiQ staff from other sites were employed on some contracts to ensure QinetiQ's offering utilised the best Pan-QinetiQ expertise. The customers were still in the main various parts of MOD, but also included defence contractors

**The Sonobuoy Sensor Systems team** offered expert services to support all stages of sonobuoy procurement including trials, based on experience of all the recent UK programmes. Sophisticated performance prediction models had been developed for trade-off studies to assist in defining the most cost-effective sonobuoy systems and to minimise the need to conduct expensive time consuming sea trials. For a while QinetiQ retained its previous DERA role as the Technical Authority for all UK procured sonobuoys, but this finished when MOD and Ultra Sonobuoys partnering relationship began, and Ultra became its own specification authority for new buoy types.

**The Maritime Airborne Processing team** offered extensive expertise and working prototypes of Multi-Sensor geographic detection displays. These had been developed to investigate fusion of the potentially vast amounts of data available in maritime scenarios, originally in support of the SR(SA)903 multistatics programme. The Maritime Airborne Processing team also offered expert services to support all stages of airborne acoustic processing and displays development including:

- Acoustic processing systems requirements capture
- ASW recording and replay systems requirements capture
- Specification writing and bid assessment

The picture below shows key members of QinetiQ's Sonobuoy Support team in 2002 with their Nimrod MRA4 IPT customer John Gillett, who for a number of years had been responsible for the procurement of sonobuoy systems for both the RAF and the RN.

Peter Martinson, Mark Brown, John Gillett, Clive Radley, and Mike Clapp

The occasion was the placing of a new one year Sonobuoy Project Support Contract with QinetiQ by John Gillett, the first one since the DERA/Dstl split and the first fixed price contract after many year of cost plus. Arrangements were made to subcontract on an occasional basis some of the DERA staff who had moved across to Dstl and in particular, Ray Cyphus, DERA's active sonobuoy expert of long standing.

# CHAPTER 13

## THE NIMROD MRA4 SAGA

By the early 1990s the Nimrod MR2 was showing its age and in 1993 a study took place to determine the outline specification for a replacement UK MPA. In 1993 a specification was issued for a Replacement MPA (RMPA) known as SR(A)420 (Staff Requirement Air 420). For a while MOD favoured the purchase of a new version of the Lockheed P-3 Orion, known as the P-7. The US Navy at that time intended to purchase the P-7, an updated P-3, but changed their minds and cancelled the P-7. A competitive tendering phase between potential contractors began in Jan 1995 and then, on 2 Dec 1996, BAE Systems were awarded a fixed-price contract as sole prime contractor, design authority and platform & systems integrator for the new aircraft – known then as Nimrod 2000.

The mid 1990s when I joined DRA Farnborough was the period when Michael Portillo the then Secretary of State Defence made the decision to place the contract for the UK's replacement MPA with BAe which resulted in the development of the ill-fated Nimrod MRA4. There were two other leading contenders in the competition both based around the turboprop powered Lockheed P3, another old airliner airframe design from the early 50s.

My impression is the decision went to Nimrod as it was thought that it had a higher UK job content than the P-3 based offerings. It is no coincidence that the three leading contenders were all based on old airframe designs as the thinking then was that this was a low cost low risk approach which proved in the case of Nimrod to be entirely erroneous. However, the Nimrod was differentiated by being jet powered, which gave it performance advantages, and it could benefit (in principle) from legacy infrastructure, which would also have been significant factors.

The first Nimrod, the MR1 was designed and developed by Hawker Siddeley who had absorbed De Havilland in 1960 the company who designed the Comet. De Havilland had its headquarters at Hatfield and lost its separate identity in 1963 and was then merged into British Aerospace in 1978 as part of Hawker Siddeley. The Hatfield site closed in 1993.

Starting in 1975, 35 Mk1 aircraft were upgraded to MR2 standard, being re-delivered from August 1979 The upgrade included extensive modernisation of the aircraft's electronic suite. Changes included the replacement of the obsolete ASV Mk 21 radar used by the Shackleton and Nimrod MR1 with the new EMI Searchwater radar, a new acoustic processor (GEC-Marconi AQS-901) capable of handling more modern sonobuoys, a new mission data recorder (Hanbush) and a new Electronics Suite.

Woodford became part of BAE Systems as a result of the £7.7 billion merger of British Aerospace (BAe) and Marconi Electronic Systems (MES) in November 1999. The aerodrome and factory became known as BAE Systems Woodford until it was sold in late 2011.

So by the time BAE took part in the competition to replace the Nimrod MR2 in the mid 1990s almost 30 years had passed since the MK1 had been designed and built. By then the corporate memory of the Mk1 design and build had largely disappeared with a lot of the original team having retired or moved on to other jobs and with numerous company reorganisations taking their toll.

If the corporate memory had been retained then the prime contractor's bid team would have been briefed during the bid stage that the old MR2 wings and fuselages were constructed before the days of CAD / CAM and were built in jigs and then mated by tradesmen hammering & filing the metal to fit as necessary – as a result each fuselage was slightly different, in some cases by up to 4 inches. Thus the embarrassing failure of the new wing designed on a modern CAD and manufactured with great precision to fit to the old fuselage wouldn't have happened. The BAE team at Woodford then found that the new wing was flawed, which resulted in the project being put on hold while another wing design was developed.

The MRA4 would have been almost a brand new aircraft, as the only the parts of the old MR2 fuselage that were being refurbished and retained were the fuselage pressure cell and empennage – everything else including the cabin pressure floor, bomb bay area, wings and undercarriage were newly designed and manufactured. Speaking with the benefit of hindsight a complete rebuild would have avoided these problems.

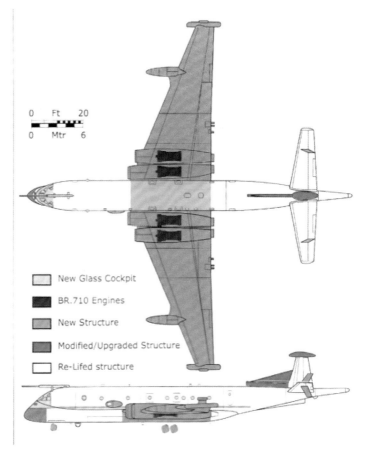

Nimrod MRA4 new build areas    Nimrods Nemesis

The following article about the RMPA competition was published in May 1996 not long before the winner was announced. It is reproduced here as it contains some interesting statements from the main bidders

shttps://www.flightglobal15 May, 1996.

**It should be remembered that the bidder's statements are essentially briefings aimed at influencing the bid outcome. So not aimed at those people doing the technical assessment of the bids but the top MOD team who were advising Michael Portillo the minister concerned.**

*With a July decision date, the RAF's Nimrod competition is reaching boiling point.*

Douglas Barrie/LONDON Graham Warwick/ATLANTA

MORE THAN 20 YEARS AGO, Hawker Siddeley emerged victorious from the ruck of the Royal Air Force's last maritime-patrol-aircraft (MPA) competition. The surprise winner of Operational Requirement 381 (OR381) was its HS.801, now better known as the Nimrod.

The surprise in the February 1965 announcement was that OR381 was a redraft of a previous MPA requirement, which appeared to observers to have been written around the Breguet 1150 Atlantic. The French offering was considered the clear favourite. The details of that decision are now buried in the murk and mythology of many a procurement saga.

The lesson will not have been lost on Lockheed Martin. The US company is now battling for the RAF's Staff Requirement (Air) 420 for a Replacement Maritime Patrol Aircraft (RMPA), with the Orion 2000, along with British Aerospace, offering the Nimrod 2000, and Lockheed Tactical System UK (formerly Loral), with its Valkyrie upgrade of the Lockheed Martin P-3 Orion.

When SR(A)909 for an acoustic processor, and SR(A)910 for a central tactical system to upgrade the Nimrod, were ditched in 1990, the clear leader to meet the RAF's replacement requirement was Lockheed's P-7, being developed for the US Navy. The axe was shortly to fall on the P-7, in part because of problems with the Boeing-developed mission-systems suite, dubbed the Update IV.

The UK Ministry of Defence, having decided against a Nimrod avionics upgrade, then set about establishing the cardinal-points specification (CPS) for SR(A)420. There are those who suggest that the latter was drafted with more than an eye on the Orion.

A request for information was released in 1993, with three aircraft quickly emerging as favourites: The Nimrod, the P-3 Orion and the twin-turboprop Dassault Atlantic.

Dassault withdrew its Atlantic 3 bid earlier in 1996, following "gentle hints" from the UK MOD that its business interests might be better served by curtailing its bid expenses for the RMPA. The RAF is firmly wed to a four-engined aircraft in the maritime-patrol role.

Since 1986, BAe has been mulling over its options for meeting a full Nimrod-replacement project and has carried out a series of studies, some at the behest of the MOD. Graham Chisnall, BAe's Nimrod 2000 project manager, says that several options were explored, covering "new designs, civil conversions and the extant platform". Civil conversions included an MPA derivative of the Airbus A310.

CIVIL CONVERSION TOO EXPENSIVE

A 25-aircraft requirement failed to merit a new-design programme, while a civil conversion was deemed to be "expensive" and "not adaptable". BAe therefore settled on the existing Nimrod platform as the basis for its bid.

The company's conclusion was that the "...most competitive bid we could offer for the RMPA competition was the Nimrod", albeit remanufactured to meet the CPS.

If settling on an airframe was a relatively straightforward process, finding a "partner" for the mission-systems area proved considerably more difficult. Alongside Boeing, BAe also held discussions with Control Data, GEC-Marconi and Paramax.

BAe was at one point struggling to pull both Boeing and GEC-Marconi into its team. Locking up the UK's other major defence manufacturer within its own bid would have strengthened BAe's hand considerably.

In the event, the three companies were unable to resolve workshare issues and GEC went its own way. In its negotiations with Lockheed Martin, GEC had looked at acting as the prime contractor, but finally opted to act as mission-system avionics integrator, rather than as prime, despite the MOD's enthusiasm for alternatives to platform primes. GEC's recent history as a platform prime contractor had been less than glorious; sources indicate that senior management took this into account when determining its strategy on the RMPA project.

The interest of Loral (now Lockheed Martin Tactical Systems UK) was also fanned by the MOD's keenness to have "systems houses" act as primes in competing for SR(A)420. Loral's reputation has also been enhanced by its management of the EH Industries EH101 Merlin anti-submarine-warfare-helicopter programme for the Royal Navy.

Like BAe, the company settled upon reworking an existing airframe. Its bid is based on upgrading ex-USN P-3A/Bs now in storage. It is also based on being the least expensive option for the MOD. Bill Vincent, the Valkyrie programme manager believes that his company has "the lowest acquisition cost".

Vincent is also aware that life-cycle support costs, rather than upfront expenditure, are the largest single contributor to an overall LCC.

His confidence also lends credence to the view that the P3-refurbishment bid is being used by the MOD Procurement Executive as a "stalking horse" to hold down the bid prices of the other, more expensive, contenders. To be more than just a method of cost-capping, however, the Valkyrie also had to be a credible, cost-effective, solution, capable of meeting the requirements.

The credibility of the bids from Lockheed Martin Tactical Systems UK and BAe are dependent on the bidders convincing the MOD that a remanufactured airframe does not constitute an unacceptable risk. The RPMA tender states: "The Authority intends to achieve value for money in a competitive environment by selecting the total solution that offers the optimum balance between the overall effectiveness, the LCC and minimised risk to the Authority."

Although the MOD had set out with the halcyon aim of procuring a non-developmental aircraft to fulfil SR(A)420, the bids received have inevitably had to include a development phase. Factored in to its analysis of aircraft LCCs, as opposed to prime bid costs, is an additional risk cost. As well as the risk associated with any given development, the MOD is also having to examine potential risks associated with refurbished airframes. It does not want any "nasty airframe surprises" waiting for it in the first decade of the next century.

BAEs Chisnall is adamant that a reworked Nimrod airframe can easily meet the 25-year lifetime stipulated in the CPS. "We have carried out an aggressive internal programme to assure the viability of the airframe and have demonstrated clearly that the retained airframe items are low risk," he says.

Two service Nimrods have been torn down to check the state of the airframes, along with the data gained from the work on airframe XV147, the form-and-fit demonstrator at BAe Warton. The results have convinced BAe that the Nimrod is good for another 35-40 years of service life with the intended airframe-modification package. "We were not talking about a marginal structure," says Chisnall.

"BEST-UNDERSTOOD RAF AIRFRAME"

BAe Chadderton has been the support authority for the Nimrod throughout the aircraft's life, and the company regards it as "probably the best-understood airframe in the RAF inventory". The effect of all this was to provide the necessary confidence for it to categorise the risk associated with the airframe as "low".

As BAe will have spent around £25 million on its bid for SR(A)420 by the time of a decision, its estimation of the state of the Nimrod airframe, on which its money is riding, cannot be dismissed lightly. It remains to be seen, however, whether the MOD will be comfortable extending for a further 25 years the life of an aircraft ordered in 1965.

As well as structural upgrades to the Nimrod airframe, the inner-wing section, wing box, elements of the vertical stabiliser and the cabin/cockpit of the Nimrod 2000 are new. BAe is also in the final throes of selecting a power plant to replace the Rolls-Royce Spey turbofans.

BAe had been expected to offer the BMW R-R BR715, but it has also been in prolonged negotiations with General Electric over a growth variant of the CF34. The engine change was driven by the need to meet time-on-station requirements and LCC criteria. If BAe had stuck with the Spey, says Chisnall, it would not have needed to replace the inner wing and carry-through structure. "This could have done another 25 years," he claims, but that boast is unlikely to be put to the test. The MOD is unlikely to opt for a Spey-powered aircraft, given maintenance and performance penalties incurred by a further 25 years of service.

Similar fears need to be allayed if the MOD is to plump for the Valkyrie. Unlike the Nimrod, however, the P-3 airframe is a relatively unknown quantity to the UK MOD, and so the likely risk element factored in its assessment of the Lockheed Tactical System UK bid will be higher.

The bid has also been clouded recently by a public quarrel with the former Loral's new owner, Lockheed Martin, over the availability of airframe data. It claims that all the necessary data were in the public domain, but Lockheed Martin argues that much of it remains proprietary to the airframe manufacturer.

The company has given the MOD assurances that both the new-build Lockheed 2000 and the ex-USN P-3s bids will "...remain on the table". Irrespective of this, Lockheed Martin is certain to want to examine the detail of its newly acquired business to ensure that it is not now bidding a hostage to fortune.

"WHAT IS GOING ON?"

"Go out and tell me what is going on, and in greater and greater detail," is how Rich Kirtland, vice-president for Government requirements at Lockheed Martin Aeronautical Systems in Marietta, Georgia, sums up the MPA mission.

Kirtland says that Lockheed Martin has looked at refurbishing the P-3 three times, first for the ill-fated P-7 programme, then for the P-7 follow-on and, finally, for the UK's RMPA competition. In each case, he says, analyses showed that a new airframe would reduce the risk. The result is the Orion 2000.

Compared with an upgraded P-3C, the new-build Orion 2000 offers a 3.5h longer endurance, and a 10% higher payload. The Orion 2000 has the same "footprint" as that of the P-3C, Kirtland says, but is redesigned to increase zero-fuel weight by 2,180kg.

The ability to increase payload weight by refining the original stress calculations and machining the structure to add or remove material is one advantage which Lockheed Martin claims for the Orion 2000 over a remanufactured P-3.

"We are talking about changes in metal thickness of thousandths of an inch. When you remanufacture an old airframe you have to bolt on plates to increase strength, and that adds weight," he adds.

The new-build Orion 2000 will also have improved material properties and manufacturing processes which increase corrosion protection. "You can manufacture-in significant corrosion resistance. Every rivet will be installed wet in the Orion 2000. We did not do that before and the P-3 corrodes around the rivet joints," Kirtland says.

A new-build aircraft, Lockheed Martin argues, also favours producing a consistent product at a consistent price. With remanufacturing, the airframes are not in a consistent condition.
The Orion 2000 and the Valkyrie are both re-engined aircraft, with Allison AE2100 turboprops. The RAF's Lockheed Martin's C-130J Hercules 2 also have this engine, and both bidders argue that the commonality offers clear cost advantages.

"For a high-/low-altitude mission profile, turboprops are very efficient, and modern commercial-turboprop technology and efficiency is phenomenal," argues Kirtland.

BAe contends that the turbofan-powered Nimrod 2000 offers lower noise and vibration, as well as a less-conspicuous power plant radar return. Not surprisingly, those offering turboprop solutions claim that, with the latest-generation engines, the noise/vibration differences are inconsequential.

BAe's campaign claims include "wrapping itself in the Union Jack", boasting that it is "the British RMPA solution". It is the UK airframe RMPA solution, but GEC, in the shape of Brian Tucker, managing director of GEC-Marconi Aerospace Systems, contends that the critical technology is not in the airframe but in the on-board avionics. Even senior BAe officials describe the airframe as "weather-proofing for the mission systems".

Boeing provides the tactical command sub-system on the Nimrod 2000, drawn from its experience on the Update IV programme for the P-7 and the Indonesian Surveiller programme. BAe will act as mission-system integrator. On the Orion 2000, GEC will provide the tactical command system (TCS), based on the architecture developed for the Nautis naval TCS, and it will act as the system integrator.

## STREET FIGHTING

Absent so far has been the shrill, near-hysterical, lobbying which accompanied the C-130J/ Future Large Aircraft (FLA) clash between Lockheed Martin and BAe. The MOD had let it be known that it did not want the high-profile, lowest-common-denominator, street fighting which characterised the RAF's transport and the Army Air Corps' attack-helicopter purchases.

As the number of airframes to be procured is limited, the "strategic-to-the-company" argument deployed by BAe in support of the FLA would court incredulity. Senior company officials argue that its importance in part lies in the fact that it acts as a gap-filler between core combat-aircraft programmes, allowing essential design-engineering teams to be retained.

One area where the procurement xenophobia did threaten was in the choice of radars. All three bidders are offering the Israeli Elta 2022 as the baseline, with the Racal Thorn Searchwater II as a costed option. The latter has attempted to lever its own case by flying the politically sensitive flag of "UK-jobs at risk". Not awarding the radar bid to Racal Thorn could also be used as a prompt to further rationalisation within the UK's radar manufacturing base. Elta has formed a bid-independent team with GEC-Marconi. Israel's Elisra is also a potential provider of electronic-support measures (ESM), marking the country's increasing interest in doing business in the UK.

BAe also argues, although its competitors find the logic questionable, that a Nimrod 2000 victory would give it an entry into the worldwide MPA market.
"There is no market for the Nimrod outside the UK. What can be exported? The Boeing mission system would have to be redesigned for an Atlantic or P-3, and that adds to the cost of the air vehicle," says Kirtland.

Not surprisingly, each of the P-3 vendors makes a strong export argument, supported by their UK associates (GEC on the Orion 2000 and Marshall Aerospace on the Valkyrie).

Lockheed Martin Tactical Systems (UK) argues that, if it were to win, and with over 75 P-3s now in storage, it would be strongly placed to pick up the bulk of procurement orders from Brazil, Germany, Italy, Saudi Arabia, South Africa and Taiwan. The UK industrial participants, the US company claims, would reap some 40% of the value of export orders.

A similar, if potentially more credible, line is pushed by its recently acquired parent on the Orion 2000. Perhaps the trump industrial card for the Orion 2000 team is the USN.

## "DEVIL IN THE DETAIL"

The USN is interested, Kirtland says, and "...is waiting to see what the UK decision is". The US defence giant is striving to get the USN to add its support to the UK bid by producing a supporting statement on the Orion 2000. Competitors claim that, should such a statement emerge, then "...the devil would be in the detail, if indeed there is any."

Lockheed Martin has already proposed that the Navy buy some 170 Orion 2000s, beginning around 2002, to avoid the cost of extending the service lives of some 240 P-3Cs and of developing a new MPA, the MP-X, to replace the P-3, beginning in 2015.

All of the remaining bidders are promising a minimum 100% offset package, while also pushing the direct-content element of their packages. Under Lockheed Martin's Orion 2000 offer, 87% of the air vehicle and 55% of the total aircraft would be manufactured in the UK.

The aircraft would be assembled at Lockheed Martin from UK-manufactured subassemblies and flown to the UK for installation of the GEC-Marconi mission system and final completion by Hunting. "The UK content in the green aircraft is guaranteed," says Kirtland.

"If the US Navy buys the Orion 2000, the same UK subcontractors will be involved," he says. While the US Navy, for example, would be likely to plug its own sensor suite into the embedded GEC-Marconi mission system, "...the UK sensor suite would be available to Orion 2000 customers without risk, and with demonstrated performance. There is more UK job content in the Orion 2000 than any other [RMPA] proposal," Kirtland asserts.

The Lockheed Martin Tactical Systems (UK) bid promises that 70% of the contract value will be bedded directly in UK firms participating on the programme. BAe says that its bid is built around some 76-80% of the work being direct, although this figure includes a substantial element of "RMPA work into the UK, via the overseas subcontractors".

"LIES, DAMNED LIES"

As with all figures, the "lies, damned lies and statistics" warning should also come into play. What should be clear, however, is that the UK industrial-participation requirement is being taken seriously by all the contenders, and that the UK aerospace industry will benefit, to a considerable degree, whichever is the winner. What the MOD may use as a discriminator in this area is each team's export potential and the likely UK involvement. Lockheed Martin and GEC are confident that the Orion 2000 export story is more sustainable than those of either of the other two competitors. BAe certainly would struggle to convince many that there is an export market for remanufactured, or even new-build, Nimrods. It is also not in Lockheed Martin's medium- to long-term export interests to see the erstwhile Loral bid win in the UK. It would far rather see new-build aircraft roll off the line at Marietta, Georgia. Lockheed Martin moved the production line to Marietta, re-opening it for South Korean aircraft.

Be it new-build aircraft or remanufactured, there is probably only a limited amount of "daylight" between at least Lockheed and BAe in mission performance. Where there may be more by way of potential discriminators for the decision makers is in risk-assessment and export.

All three contenders would claim that the above criteria would see them emerge as the winner. Lockheed Martin would, however, be more concerned about whether a high-level "British Aerospace-jobs-equals-votes" taint enters the decision.

**Lockheed Martin's assessment was correct and BAE got the contract.**

| Prototypes | | | |
|---|---|---|---|
| PA01 | ZJ516 | First flight 26 August 2004 | (ex XV247) |
| PA02 | ZJ518 | First flight 15 December 2004 | (ex XV234) |
| PA03 | ZJ517 | First flight 29 August 2005 | (ex XV242)* |
| Production | | | |
| PA04 | ZJ514 'A' | First flight 11 September 2009 | (ex XV251) |
| PA05 | ZJ515 'B' | First flight 8 March 2010 | (ex XV258) |
| PA06 | ZJ519 'C' | | (ex XV284) |
| PA07 | ZJ520 | | (ex XV233) |
| PA08 | ZJ521 | | (ex XV227) |
| PA09 | ZJ522 | | (ex XV245) |
| PA13 | ZJ517 | | (ex XV242) |
| PA10 | ZJ523 | | (ex XV228) |
| PA11 | ZJ524 | | (ex XV243) |

Nimrod MRA4 Flight/Build Programme

After the Nimrod MRA4 programme was cancelled by Liam Fox various experts gave evidence to the Parliamentary Defence Sub Committee. One of these was Dr Sue Robertson who had been a subject matter expert in the MRA4 Mission System team.

# THE MRA4 CANCELLATION DECISION

By the early 1990s the Nimrod MR2 was showing its age and in 1993 a study took place to determine the outline specification for a replacement UK MPA. In 1993 a specification was issued for a Replacement MPA (RMPA) known as SR(A)420 (Staff Requirement Air 420). For

The following is my summary of written evidence from Dr Sue Robertson to the Parliamentary Defence Committee re the UK's MPA situation in 2011. She was until October 2010 the Subject Matter Expert on the Electronic Support Measures (ESM) system for the Nimrod MRA4 and worked, on behalf of the Ministry of Defence, on the evaluation of the system and advised on changes to the system. Her previous role had been to carry out the same function on the Merlin Mk1 helicopter programme.

I have included this because if anyone is in a position to know what really happened, Dr Robertson should. However, parts of it are opinion, an expert opinion, but not fact.
But importantly Dr Robertson was not employed by BAE and has no affiliation with the RAF, the Navy or the Army.

**Dr Robertsons' Evidence**

Just after the announcement that the Nimrod MRA4 programme was to be cancelled in October 2010 I wrote to the Prime Minister pointing out that none of the primary roles of the MRA4 could be carried out adequately by un-manned vehicles or by satellite surveillance, due to the complex nature of the electronic and acoustic signals which must be interpreted by highly experienced Operators with knowledge of the fine detail of the signals that they are observing. The other roles of the aircraft such as long-range maritime search and rescue, terrorist threat interception and disaster response co-ordination also re The MRA4 was not just a submarine-hunter, it was capable of a variety of roles from ship surveillance to search and rescue which quire human intervention in real-time.

I did not even consider the possibility that the MOD would try to use Merlin Helicopters, Type 23 Frigates and C130 aircraft to fulfil the roles of the MRA4. Much to my surprise the response from the MOD has been that they are planning to use these platforms, none of which acting alone or together can provide an adequate substitute for the Nimrod.

The following table shows the roles which the MRA4 would have carried out and the capability of each of the alternative platforms against each of the roles.

**Table 1**

*The Roles of the Nimrod*

| Asset Task | Nimrod MRA4 | Merlin Mk1 | Type 23 | C-130 |
|---|---|---|---|---|
| Submarine Detection (ASW) | Yes - 6000Nm (nautical mile) range with 15-hour mission time | Yes - 200 Nm (nautical Mile) range with 90-minute mission time | Yes | No |

| Shipping Surveillance | Yes - to 260Nm (nautical mile) at 40,000 ft | Limited Sensors | No | Limited - no adequate sensors |
|---|---|---|---|---|
| Fleet Protection | Yes | Yes | Limited range | Limited - no adequate sensors |
| ISTAR (Support of Troops in Afghanistan) | Yes | No | No | No |
| ELINT Data Gathering | Yes | No | No | No |
| Counter-terrorism | Yes | No | No | Perhaps |
| Weapons Deployment | Yes | Yes | Yes | Yes? |
| Search & Rescue | Yes - 2400Nm (nautical mile) range for three hours search | Limited - 300Nm (nautical mile) range with one-hour search | No | Limited -600Nm (nautical mile) range with two hours search |
| Emergency Communications | Yes | No | No | Yes |
| Overseas Maritime Patrol | Yes | No | No | No |
| Counter-pirate operations | Yes | No | No | No |
| Protection of Trident Submarines | Yes | Limited range | Limited range | No |
| Protection of Future Carriers? | Yes | Limited range | Limited range | No |

One of the most important roles of the Nimrod MRA4 would have been as an ISTAR (Intelligence, Surveillance, Target Acquisition and Reconnaissance) and ELINT (Electronic Intelligence) data gathering platform. The aircraft could perform long-range communication monitoring and image capture. The ESM system was capable of detecting, locating and identifying radar signals (from fixed sites and moving platforms) at a range of over 200 nautical miles. The fusion of data from multiple sensors meant that an excellent tactical picture of the operational environment could be built up, a vital requirement in support of current operations in Afghanistan. The very capable recording equipment on the aircraft meant that valuable ELINT data could be made available for detailed post-flight processing by the Air Warfare Centre to enable the population of strategically important databases and the projection of future threat trends.

It could act as a communications and disaster co-ordination platform and perhaps its most important role would have been as an ISTAR (Intelligence, Surveillance, Target Acquisition and Reconnaissance) platform in support of operations in Afghanistan.

The production of an accurate tactical picture could have contributed hugely to the safety of our troops. We were about to undergo a step-change in the quality and quantity of Electronic Intelligence (ELINT) data recorded by the MRA4 for the population of strategically important databases for the Air Warfare Centre. A key component of the UK ISTAR capability has been lost with the demise of the MR ISTAR and ELINT Data Gathering

During MRA4 Mission Systems tests it had already been demonstrated that the capability of the MRA4 to carry out ELINT tasks far out-shone that seen on any of the UK other ISTAR platforms.

Other aircraft operated by the UK, the AWACs E-3 and Nimrod MR2 shared the same type of ESM system, which was not good enough for serious ELINT data gathering and they had limited recording facilities. The Nimrod R1, which is about to be retired, had limitations in coverage for Electronic signal processing. The UK is to take delivery of 3 Rivet Joint aircraft in 2014, re-fitted (not re-built) Boeing 707 aircraft which are over 40 years old and are to be operated as joint fleet with the US—they will probably provide us with some ELINT data.

However, with the introduction into service of the MRA4 and its subsequent wider operational coverage, the UK was about to undergo a step-change in the quality and quantity of ELINT data which would have been available.

A key component of the UK ISTAR capability has been thrown away with the cancellation of the MRA4 programme.

## SEARCH AND RESCUE

As a nation we are now failing in our international search and rescue obligations. The region for which the UK is responsible extends from 45 to 61 degrees North and from 3 degrees East to 30 degrees West. This means the UK should be able to offer assistance to vessels which are up to 1,200 nautical miles from our coast. to 30° West. The Merlin has an effective search and rescue range of 300 Nm (nautical mile) (with one-hour search) and the C130J has a range of 600Nm (nautical mile) (with two hours search). The C130J does not currently have adequate sensors to perform efficient search and rescue operations. Neither platform can be effective for long-range search and rescue as the endurance of these aircraft is insufficient to allow for search patterns of useful duration to be carried out. We do not have a Search and Rescue aircraft with the range and endurance for this task. MRA4 had a 6,000 Nm (nautical mile) range and a 15-hour potential mission time, so it would have been able to fulfil our international obligations for search and rescue.

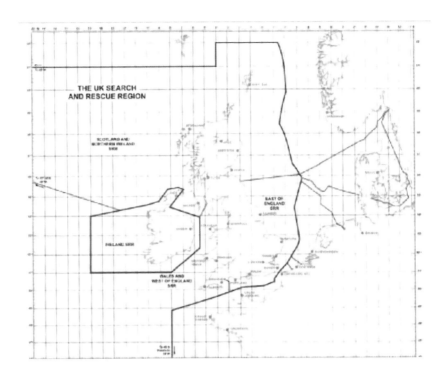

# THE TRADITIONAL THREAT- SUBMARINES - THE ASW ROLE OF NIMROD

Over 40 countries have submarines in service and many, such as China, North Korea and Iran are still building them. Here are some examples of the numbers of submarines which are currently in service.

| | |
|---|---|
| **Iran** (21) | **N Korea** (70) |
| **China** (>70) | **Egypt** (8) |
| **Pakistan** (5) | **Russia** (> 80) |
| **Argentina** (3) | **Algeria** (4) |

It was reported in the Daily Telegraph (27 August 2010) that Russian submarine activity around UK waters had reached levels not seen since 1987. Russian Akula submarines were attempting to track Vanguard class submarines which carry the UK nuclear deterrent. It is understood that the Russians stood off Faslane, where the British nuclear force is based, and waited for a Trident-carrying boat to come out for its three-month patrol to provide the Continuous at Sea Deterrent.

Within days of the cancellation of the MRA4 there were two more publicly acknowledged "submarine incidents":

1. The submarine "Astute" went aground in full view of any ship, foreign submarine or aircraft who cared to look and we have no idea who was looking!

2. A Russian submarine was "lost" during an exercise involving Nato aircraft. The Akula submarine disappeared after being sighted in the North Sea. Two US Orion P3 aircraft which were taking part in the Nato Joint Warrior exercise tried to find the submarine, but failed to locate it.

Here is what a Royal United Services Institute analysis report (by Lee Willett, January 2011) has to say about the loss of the MRA4 ASW capability:

"The submarine threat is a significant national security issue, not just a Cold Warrior's hangover."

"Despite MOD statements that Nimrod's roles will be covered by other assets, no other assets deliver its specific capabilities. The UK's ASW web hence has a particular, and significant, hole in it."

"In Nimrod, the refined sensor capabilities—both actual in the MR2 and planned in the MRA4—together with the aircraft's range, speed and endurance, gave the UK an asset which could operate from strategic to tactical levels. Operating in all three environments—air, surface and sub-surface—it could reach targets, even distant ones, quickly and could maintain pressure on the target while vectoring in other assets."

"The Type 23/Merlin package does not match Nimrod's capability."

Here is what the National Audit Office Report HC489-I, Session 2010-11 published on 15 October 2010 stated about capability risks:

"Loss of the capability offered by the Nimrod Maritime Reconnaissance and Attack Mk4 would have an adverse effect on the protection of the strategic nuclear deterrent, the provision of which is one of the Ministry of Defence's Standing Strategic Tasks. In addition, the maintenance of the integrity of the UK through detection of hostile air and sea craft would be compromised."

## SHIP SURVEILLANCE

The increase in the possibility of terrorist attacks means that we need to protect our shores more now than at any time in recent years. The platforms which have been proposed to carry out the MRA4 roles do not have the coverage to be able to effectively monitor shipping around the UK. The Nimrod sensor operators are experts at recognising patterns in shipping movements and the sensor range of the MRA4 would have given the best chance of identifying and countering threats.

The skills of the excellent RAF sensor operators who undertook the testing of the aircraft and who were to be the instructors of the MRA4 operational crews are shortly to be lost. They are expecting to be made redundant later this year. The interpretation of Sensor data is complex and it takes years of practice to be able to extract the important features from a complicated picture.

The Merlin helicopter has an ESM system that is not capable of producing an accurate picture of the electromagnetic environment when operated over-land or in littoral waters.

# THE REASONS BEHIND THE MRA4 CANCELLATION DECISION

When the Nimrod MRA4 programme was cancelled in October 2010 the Government stated that the decision had been taken to save money. Although many people were worried about the lack of Maritime Reconnaissance, the need to cut back on all areas of government spending meant that the UK may have had to accept that we would be without effective Maritime Surveillance for the time being. I, like most people, assumed that the aircraft would not be completely destroyed until further consideration had been given to the capability loss and that it should be possible to "mothball" the aircraft in case the requirement changed in the future. The contract that MOD had with BAE Systems was terminated "for the Convenience of the Customer".

However, despite appeals from many people, the Government ordered the immediate destruction of the aircraft. The considerable opposition to this action led to claims by the Defence Secretary that the aircraft had not flown, that the aircraft was unsafe, it was more than 10 years late and that those working on the aircraft did not believe that the "technical difficulties" with the aircraft could be overcome. The decision to destroy the aircraft before the Defence Select Committee had reviewed the SDSR led to protests from many different groups of people. The Government then tried to give other excuses for the cancellation of the programme.

I have provided information in the section "In Defence of the MRA4" to refute the government allegations that the aircraft had not flown, that it was unsafe, that it was 10 years late and that those working on the programme did not believe that the technical difficulties could be solved.

There are four issues here:
1.  The implication that the Nimrod had not flown.
2.  The assertion that MRA4 was unsafe.
3.  The claim that the Nimrod was 10 years late.
4.  The implication that those working on the aircraft did not believe it was viable.

**The Nimrod had not passed its flight tests yet"—quote by the Secretary of State for Defence in a BBC TV broadcast on 27 January 2011**

Here is a summary of the flight history of the five MRA4 aircraft that had been flying.

PA01 first flew on 26 August 2004. It had no mission system, but was used for air-frame testing for the next five years.

PA02 first flew on 12 December 2004. It was used extensively for Mission System Testing and had completed over 230 flights, including testing in the McKinley Climatic Facility at Eglin Air Force Base in Florida and airfield performance trials at Asters in France.

PA03 first flew on 29 December 2005. It was used for Mission System Testing and had completed over 60 flights.

PA04 first flew on 10 September 2009. It was delivered to, and accepted by, the RAF on 19 March 2010. At the time of cancellation of the programme the aircraft was cleared for flight and had recently been flown by BAE Systems personnel.

PA05 first flew on 8 March 2010. Mission system data was recorded during flights of this aircraft.

The original plan was for PA01, 02 and 03 not to be used in service, however, it should have been possible for PA02 and PA03 to have been put into operation with relatively little extra cost, so there would have been four aircraft ready for immediate use.

**The MRA4 was "Unsafe"**

A document "leaked" to The Sunday Times led to the following newspaper article: "MOD Documents reveal that Nimrods had 'critical' fault" By Simon McGee in The Sunday Times on 30 January 2011

In the following paragraphs the text in italic font indicates the text of the newspaper article. I have interspersed the text with comments on each of the "issues" which are mentioned in the article.

*The nine Nimrod aircraft cancelled amid a storm of condemnation and at a cost of £4 billion were designed with the same critical safety fault blamed for the downing of an RAF Nimrod in 2006 with the loss of 14 lives. Liam Fox, the defence secretary, has been accused of leaving a "massive gap" in the nation's security by scrapping the fleet of maritime patrol planes. But classified documents seen by The Sunday Times reveal Ministry of Defence (MOD) safety tests conducted last year on the first Nimrod MRA4, built by BAE Systems, found "several hundred design non-compliances".*

There are always non-compliances which emerge through a development cycle as complex as that of a military aircraft. They are either fixed, or it is agreed that they can be accepted into the design. Either way, they remain on the "list" forever, forming part of the record of the development process.

*Problems opening and closing the bomb bay doors*
There was no problem with the bomb bay doors.

*Failures of the landing gear to deploy*
The landing gear never failed to deploy/retract. There were two instances of nose wheel door indication failure due to incorrectly positioned nose wheel door micro-switches—this had been fixed.

*Overheating engines*
There are no recollections of any engine overheating during flight trials.

*Gaps in the engine walls*
Gaps were found between the engine bay fire wall and surrounding structure. A temporary fix enabled flight test to continue and a permanent fix was later found.

*Limitations operating in icy conditions*
There were limitations at the time of cancellation of the programme as QinetiQ had not finished its final testing. As QinetiQ provided wider clearances the limits on operating conditions would have been expanded.

*Concerns that "a single bird-strike" could disable the aircraft's controls*
3.3.9 If a bird had flown into the open bomb bay and hit an area 6 inches' x 4 inches there may have been an effect on the aileron system. A cover guard was to be introduced to mitigate against this remote possibility.

*The most serious problem discovered by Defence Equipment and Support (DE&S) inspectors at MOD Abbey Wood in Bristol involved a still unresolved design flaw. It concerns the proximity of a hot air pipe to an uninsulated fuel line, widely blamed for an explosion on board Nimrod XV230 on 2 September 2006, near Kandahar airport in Afghanistan. A three-page summary of the faults, labelled "restricted" and written on 17 September, last year, stated: "The work being undertaken by the MOD to validate the BAE Systems aircraft's safety case during the week of 13 September 2010, identified a potentially serious design defect: a small section of a hot air pipe was discovered to be uninsulated in an area that also contains fuel pipes, which is outside the design regulations."*

All the control, fuel, engine and mission systems on the MRA4 were new design, the only parts carried over from the MR2 were the fuselage and empennage. The hot air pipe which is referred to in the newspaper article was insulated along its length of more than 8ft, apart from the last 4 inches, where it went through a bulkhead into the intake nacelle. The fuel feed pipe from the No. 1 tank travels through the same space between the fuselage and the inboard engine fire wall. Analysis looked at the likely failure rate of the fuel pipe, the maximum temperature of the engine intake anti-icing off-take and the likely usage time of the engine anti-icing system and concluded that a design solution would be needed. This was being worked on and a temporary solution was to isolate the No. 1 tank which would have resulted in a temporary reduction in flight duration. The loss of XV230 in Afghanistan was caused by fuel leakage into the bomb bay.

At the time that the programme was cancelled the MRA4 was cleared for flight by BAE Systems flight crew, but the process of flight clearance for RAF crews had not been completed by the Military Aviation Authority (MAA). The MRA4 was the first aircraft to undergo the process of release-to-service by the newly-established MAA.

**Length of Programme - Quote by Dr Liam Fox "The programme was 10 years late"**

The initial contract for the development of the MRA4 was signed by the then Conservative government in December 1996. The in-service date for the aircraft fleet was slipped (to the surprise of those of us working on the project) last year by a year to autumn 2011. This would have given a total development time of 14 years. Whilst not ideal, this does not compare unfavourably with other aircraft development programmes, as shown in the following table.

**Table 2**

*Aircraft Programme Durations*

| Aircraft Type | MRA4 (UK) | Typhoon (UK) | Merlin Mk1 (UK) | F-35 (USA) | Rafael (France) | Atlantique 2 (France) | P-8 (USA) |
|---|---|---|---|---|---|---|---|
| Time from start of development to in-service | 14 years | 15 years | 18 years | 15 years | 19 years | 12 years | 9 years |

It seems that the Government has a naive view of how long it takes to develop and commission military aircraft, given the time-scales of other military aircraft programmes listed above.

If the MOD believe that the MRA4 was 10 years late, the aircraft would have been expected to have been delivered in 2000, only three years after development had begun. In fact, the original in-service date for the MRA4 was 2003, which was in itself unrealistic.

For example, the at Atlantique 2 was a re-fitted version of Atlantique 1 and took 12 years to be put into service after its re-fit.

The MRA4 is a not a re-fit of the MR2, it is a re-built aircraft and it could have met its in-service data of 2011.

*They were not even sure that they could resolve some of the technical difficulties..."quote by the Secretary of State for Defence on British Forces News (3 February 2001)*

BAE Systems are quoted as saying (The Sunday Times 30 January 2011) that the time of the cancellation of the MRA4 programme, we were working with the Ministry of Defence—in the normal way—to resolve a number of issues relating to the aircraft. We are confident that these would have been resolved to enable the aircraft's entry into service as planned. Having worked on the Mission System, I am also confident that the remaining issues would have been resolved sufficiently for the aircraft to have provided adequate capability on its entry into service.

It is unrealistic to expect that any military aircraft will be perfect in every respect at the start of its service life. I would have expected that, in common with other aircraft programmes, the optimisation of the mission system would have been an on-going activity throughout the life-cycle of the aircraft fleet.

Dr Sue Roberts - February 2011

Air Vice – Marshall Andrew Roberts submitted a similar paper titled:

### UK MARITIME PATROL AIRCRAFT – AN URGENT REQUIREMENT

This paper concentrated on the capability gap caused by the cancellation of the Nimrod MRA4 and the need in his view to restore the UKs Maritime Surveillance capability as soon as possible. The views expressed were very similar to those aspects of Dr Roberts' paper.

As far as the conflicting views expressed by Dr Roberts and the MOD on the technical issues surrounding the MRA4 cancellation decision goes, I will make no further comment as I am not an expert in that area.

However, since 2011 the government has performed a U turn and decided that the UK requires a fixed wing MPA after all.

**The Boeing P8 MMA will be patrolling the waters around the UK at some time in the future, 2020ish?**

# CHAPTER 14

## THE UK SONOBUOY/MPA SITUATION IN APRIL 2016 FOLLOWING THE P 8 PURCHASE DECISION

Nimrod MRA4 *(Photo: BAE)*

As a result of the cancellation of the Nimrod MRA4 aircraft which was to be the main UK aircraft to use sonobuoys replacing the MR2, the only current (2016) UK asset to employ sonobuoys is the Merlin Helicopter. The Merlin has limited range and endurance compared to a conventional aircraft for ASW.

Merlin *(Illustrations: Westland)*

To remedy this, the UK government announced in November 2015 that nine Boeing P 8 MPAs would be purchased.

Before 2015 UK MOD had already taken the decision to retain the skills of ex –Nimrod aircrews through what is called the Seedcorn programme and approximately 30 people have been placed on extended exchange with the USN, RCAF, RAAF and RNZAF and are operating their MPAs.

The UK's purchase of the P 8 will enable the UK in the future to obviate the need to call on its NATO allies to conduct searches with MPA aircraft equipped with advanced sensors for missions such as:

- An airliner which disappears say 500 miles west of Ireland or Scotland.

- Foreign submarines patrolling on the exit routes of our SSBNs from Faslane.

- Surface vessels of any size attempting to approach the UK covertly.

UK area of responsibility in light shade

Reproduced from Nimrods Rise and Fall *Tony Blackman*

Boeing P 8 MPA

The P 8 is being developed in three versions known as increments 1 and 2 and 3 for the USN plus the export ones for India and the Australians.

Increment 1 is not being built any more. It had some P 3 legacy electronic/acoustic processing systems from the P 3 and is already out of date compared with increment 2. Increment 2 is the current USN production version and increment 3 is the most advanced version with a sophisticated US electroacoustic Multistatic system.

In March 2016 I received a letter from Michael Fallon via my local MP.

Mr Fallon's letter announces that the UK is purchasing off the shelf increment 3 P 8s.

**11 FEB 2016**

**Ministry of Defence**

**SECRETARY OF STATE**
MINISTRY OF DEFENCE
FLOOR 5, ZONE D, MAIN BUILDING
WHITEHALL  LONDON  SW1A 2HB

Telephone 020 7218 9000
Fax: 020 721 87140
E-mail: defencesecretary-group@mod.uk

Our ref: D/S of S/MF MCSOS2016/00413e          *8th*          February 2016

*Dear Michael*

Thank you for your letter of 6 January 2016 concerning your constituent, Mr Clive Radley of 37 Buckingham Way, Frimley, Camberley and his queries on the Boeing P-8 Maritime Patrol Aircraft.

I can confirm that the Department is acquiring the P-8 aircraft as an 'off-the-shelf' procurement, which will be delivered to the same specifications and with the same modifications as a standard US Navy P-8.  We are developing plans to address the interoperability issues arising from the difference between the US sonobuoys used on the P-8 and the UK equivalents used on the Merlin. This may include software or hardware modifications and/or the use of certain tactics, techniques and procedures.

Under current plans, all UK P-8 aircraft will be built to Increment 3 standards. We will also join the development programme to upgrade the aircraft through life.  At this stage the Department has no plans to procure the same wide area surveillance UAV as the US Navy.  However the Department is continuing, through its capability development and financial processes, to examine the case for enhancing future wide area surveillance capabilities.

The Department has no plans at this stage to procure a MAD equipped Unmanned Aircraft System, but we will continue to monitor the development of such systems as part of our future capability planning.  As the UK P-8 enters service, it will be equipped with the same deployable equipment and weapons as the US Navy's P-8s.

The Government took the decision in the 2010 Strategic Defence and Security Review (SDSR10) to cancel the Nimrod MRA4 programme, which was 11 years late and nearly £800 million over budget, on the clear understanding that this would result in a gap in the UK's intelligence, surveillance and reconnaissance capabilities.  As planned at the time, this gap is being mitigated by drawing on a combination of other UK maritime platforms and other surveillance aircraft, including those from allies, as described in section 2.E.1 of SDSR10.

*Yours ever*

**THE RT HON MICHAEL FALLON MP**

Rt Hon Michael Gove MP
House of Commons
London
SW1A 0AA

**On a negative note with respect to the current (April 2016) UK capability gap in Airborne ASW and SAR etc., the increment 3 P 8 is not due in service until 2020/2021. So another five years of depending on our NATO allies.**

Reading between the lines it appears that there will be little UK supplied sonobuoy and acoustic processing content for the foreseeable future. Our initial in-service P 8s will not be able to deploy sonobuoys built in Greenford. The P 8 does not currently process UK sonobuoy information, and it is not cleared to deploy non A-size sonobuoys. This might mean our initial in service P 8s might employ only US built and specified sonobuoys. Furthermore, P 8s will deploy US built torpedoes rather than the UK Stringray torpedo.

For QinetiQ Farnborough this would mean that in the short term they may not have a role in the sonobuoy and airborne processing aspects of the P 8.

The P 8s engines are very thirsty at the normal low ASW mission altitudes previous MPAs have operated at, like the Nimrod MR2 and the P 3. This thirstiness will require changes in the way the P 8 operates compared to the P 3 and Nimrod, i.e. spend most of the time patrolling at high altitude. Press reports suggest that unmodified P 8s will not able to refuel from our new Voyager Tankers unmodified. If true, this would limit the P 8's operating radius and ferry range and time on station. Media reports say the P 8 and Voyager are incompatible and this seems to be acknowledged by MOD.

It means the P 8 will have a shorter range, and time on station, than the cancelled MRA4. This also implies that the performance etc requirements our P 8s will have to meet are different to those that the MRA4 was designed to meet, ie. less demanding.

It may be that once the P8 is in service with the UK that modifications to the P-8's ASW system will be considered, e.g. the carriage of G-sized Sonobuoys, or MSA processing, but if that happens it will be many years in the future.

Ultra has delivered multistatics processing software to Thales UK, for integration into the Merlin Mk2 Helicopter Sonics processing system. This suggests that QinetiQ Farnborough's involvement with sonobuoys is likely to continue in the immediate future in the form of providing technical advice to the Merlin project team.

## SONOBUOY MINIATURISATION FOR UNMANNED ASW

The UK's Defence Science and Technology Laboratory (Dstl) and industry partners are currently exploring the potential for Sonobuoy System Miniaturisation. This effort is driven by the likely emergence of unmanned air, surface, or underwater vehicles (UAV, USV, UUV), generically known as UXV, in the future maritime battlespace. It is believed unmanned vehicles will play an increasing role as navies move away from legacy platform-based thinking and embrace new 'Distributed ASW' operational concepts. involving remote off-board systems.

However, the size and weight of current generation sonobuoys impacts significantly on UAV payload/radius. To overcome this constraint, Ultra Electronics Sonar Systems as the major industrial partner is carrying out a three phase programme for Dstl:

Phase 1 of the Sonobuoy Miniaturisation project involved system-level studies to quantify the effects on performance, and examine integration with UAV platforms. This stage resulted in the development of a series of high-level design options for miniaturised sonobuoys and associated UAV payload pods.

During Phase 2, Ultra de-risked key technologies and built hardware prototypes for demonstration.

This outcome of this work has been to mature designs for a new generation of G-size and F-size buoys, which are approximately ½ and 1/3 of the standard NATO A-size buoy usually carried on large maritime patrol aircraft, but with equivalent or better overall system performance.

A planned Phase 3 will demonstrate the operational use of unmanned and miniaturised Sonobuoy Systems.

Phase 3 will fully demonstrate Sonobuoy System Miniaturisation.

# SUMMARY CONCLUSION RE THE 2016 AIRBORNE ASW SITUATION AT FARNBOROUGH

In conclusion, the high MOD funding levels allocated to ASW during the height of the Cold War are unlikely to be repeated. During that period RAE Farnborough scientists with the support of AUWE etc were able to:

- Conduct fundamental underwater acoustics research

- Carry out applied research on all aspects of airborne ASW from the wet end buoy, the uplink to the MPA, the receivers and airborne processing and displays

- Contract companies such as Ultra and others to produce prototypes etc. and to investigate their own novel ideas

- Conduct many trials with MPAs and friendly submarines to evaluate ideas in the real world

This activity played a key role in the Cold War and made a major contribution to the UK's world leading airborne ASW capability during that period.

This activity played a key role in the Cold War and made a major contribution to the UK's MPA's world leading ASW capability during that period.

Some of the RAE's key ideas were:

1. The early RAE designed active sonobuoy T 7725 built by Dowty (now Ultra Electronics Sonar Systems) purchased by the US as the AN/SSQ-20 as their own sonobuoy had failed
2. GPS in sonobuoys making target localisation quicker and more accurate
3. The CAMBS active localisation sonobuoy
4. Energy Map geographic sonobuoy field detection displays
5. RFS for more flexible use of uplink channels
6. Smaller sonobuoys and digital telemetry with the same performance as the standard A-size

It must be emphasised that UK Ultra worked closely with the RAE etc on all these ideas and came up with many innovative ideas in their hardware and software solutions. UK MPA crews also worked closely with the RAE and Ultra.

# BIBLIOGRAPHY

- J G Rees Author of HUMDE Portland Research Note 136 Radio Sonobuoys November 1951

- Sparton Electronics (1984 Sonobuoy paper by Russell Mason)

- Not Ready for Retirement: The Sonobuoy Reaches age 65 - November Sea Technology 2006 paper by Roger Holler, Arthur Horbach and James McEachern of Navmar Applied Sciences Corp Warminster.

- The Ears of Air ASW – A history of US Navy Sonobuoys – R. A Holler, A.W. Horbach, James F McEachern published in 2008.

- The Evolution of the Sonobuoy from World War I to World War II to the Cold War– paper by Roger A Holler Navmar Applied Sciences Corporation

- U.S. Navy Journal of Underwater Acoustics JUA_2014_018_A June 2014 Introduction to the Theme: Airborne Anti-Submarine Warfare - Peter W. Verburgt Naval Air Warfare Centre, Aircraft Division Mission Avionics Department

- Aircraft versus Submarine in two World Wars by Dr Alfred Price first published in 1973

- Aircraft versus Submarine – The evolution of anti-submarine aircraft 1912 t0 1980 by Alfred Price published in 1978 and 1980

- Seek and Strike – Sonar, Anti-Submarine Warfare and the Royal Navy 1914-1954 - Willem Hackman– published 1984

- Nimrod: Rise and Fall by Tony Blackman (Author)
- BAe Nimrod by John Chartres - Ian Allan ltd, published in 1986

- Extract from an article in  RAF Historical Society Journal 33 written by Air Cdre Bill Tyack retired. The article describes a typical Shackleton ASW exercise

- Extract from an Article in  RAF Historical Journal 33 written by Sqdn. Ldr I M Coleman. The article describes a typical Nimrod ASW mission

- The papers of Group Captain Hugh Williamson- Churchill Archives Centre, The Papers of Group Captain Hugh Williamson

- Xxxx Parliamentary defence evidence

- Nimrods Genesis Chris Gibson

- Johnathon Dimbleby- The Battle of the Atlantic

- Avro Shackleton - Owners Workshop manual Haynes

- Blackett's War -Stephen Budiansky – Vintage book s2013

- The Silent Deep – Peter Hennessy and James Jinks

- Avro Shackleton The RAFs Cold War Sub Hunter -Aeroplane Illustrated

- Hunters Killers - Iain Ballantyne

- The RAF in Camera – 1903-1939 Roy Conyers Nesbit – Alan Sutton Publishing

- The RAF in Camera – 1946-1995 Roy Conyers Nesbit – Alan Sutton Publishing

- Coastal Command in Action 1939- 1945 Roy Conyers Nesbit

- Farnborough 100 Years of Aviation - Peter J Cooper

- Flying boats of World War - AEROPLANE ILLUSTRATED

- The Growler Magazine – The Shackleton Association

- Flypast Magazine   - various months

- International Federation of Operational Research Studies(IFORS) - Operational Research Hall of Fame 2003 Patrick Maynard Stuart Blackett 2003

- Aeroplane Monthly – various months

- The RAF Yearbook 1980 and 1981 – Published by the RAF Benevolent Fund

- Coastal Command – the Air Ministry Account of the Part Played by Coastal Command in the Battle of the Seas 1939 to 1942 -Published in 1942 by His Majesties Stationer Office price two shillings

- British Airpower in the 1980s-The Royal Air Force – Air Commodore R a Mason – Published in 1985 by Book Club Associates

- Multistatic sonar: a road to maritime network enabled capability -Robert Been, Stephane Jespers, Stefano Coraluppi, Craig Carteel, Christopher Strode, Arjan Veermeij, October 2007. Originally published in UDT Europe, Naples Italy June 2007

Useful information was obtained from the following websites:

The British Newspaper Archive www.britishnewspaper archive.co.UK

The British History Online project www.british-history.ac.UK

Janus: The Papers of Group Captain Hugh Williamson
http://janus.lib.cam.ac.UK/db/node.xsp?id=EAD%2FGBR%2F0014%2FWLMN

https://scripps.ucsd.edu/labs/buckingham/wp-content/uploads/sites/60/2014/07/MJB_CV_July05.pdf

GlobalSecurity.org news@news.nl00.net
http://www.militaryaerospace.com/articles/2014/12/p8-high-altitude.html

http://www.dtic.mil/get-tr-doc/pdf?AD=ADA610348
http://www.ainonline.com/aviation-news/defense/2013-06-17/p-8a-poseidon-readied-submarine-warfare
http://tacticalmashup.com/listening-sticks-us-navy-sonobuoy-contracts/

http://www.spyflight.co.UK/mr4a.htm
http://aviation.maisons-champagne.com/dir.php?centre=04-bio-aubrun&menu=11

http://sparton.com/webfoo/wp-content/uploads/temp_file_9-06-11_Sparton_Awarded_HAASW_Development_Contract1.pdf

http://goaeis.com/Portals/GOAEIS/files/EIS/Appendices/GOA_FEIS_Appendix_H.pdf

http://www.publications.parliament.UK/pa/cm201213/cmselect/cmdfence/110/110vw05.htm

http://www.Ultra-electronics.com.au/resources/Sonobuoy%20techsystems.pdf

I have attempted to ascertain the copyright status of the photographs, etc. used in the book, but this was not possible in every case.

# APPENDIX 1
# GLOSSARY

| | |
|---|---|
| ALFEA | Active Low- Frequency Electro Acoustic |
| ASSS | Active Sonobuoy Search System, UK 1990s experimental multistatic active search system comprising transmit and receive sonobuoys and airborne processing and displays |
| ADAR | US Air Deployable Active Receive Sonobuoy.   AN/SSQ-101 |
| AEER | US Advanced Extended Echo Ranging Transmit Sonobuoy. AN/SSQ110B |
| ADLFP | Advanced Development Low Frequency Projector. US Coherent Acoustic Source Sonobuoy now (SSQ 125) |
| Anechoic coating | a coating applied to submarines to reduce the strength of echoes reflected to receiving sonobuoys and sonar |
| ASDIC | Active sonar first developed in the UK in 1918 for Anti-Submarine Warships. Anti-Submarine Division Investigative Committee |
| A-size sonobuoy | 4 and 7/8 inches' diameter, 36 inches' long |
| AUWE | Admiralty Underwater Weapons Establishment |
| ASW | Anti Submarine Warfare |
| Barra | Passive receive sonobuoy originally developed in Australia. |
| BAE | BAE systems formerly British Aerospace |
| B-size sonobuoy | 6 inches' diameter, 60 inches' long |
| BENDER | A flexural hydrophone or projector, comprising two metal plates, in which a piezoelectric ceramic disk is bonded to at least one of the metal plates, which bends due to differential strain when a voltage is applied across the ceramic element. Benders are typically used as hydrophones or as projectors in sonobuoys, often as part of an array, to provide receive or transmit directivity and array again |
| BROADBAND | Emission or detection of non-coherent acoustic energy over a wide bandwidth (e.g. acoustic emissions due to flow noise or propeller cavitation) |
| C-size sonobuoy | 9 and 1/2 inches' diameter, 60 inches' long |
| CAMBS | Command Active Multi-beam Sonobuoy – The UK's current active detection and localisation sonobuoy |
| CDA | Centre for Defence Analysis. UK MOD organisation which conducted Operational Analysis during the Cold War |
| CFS | Command function selection explain |
| CPA | Closest point of approach |

| | |
|---|---|
| DICASS | Directional Command Activated Sonobuoy. The current in service US active localisation sonobuoy. |
| DRA | Defence Research Agency |
| DERA | Defence Evaluation and Research Agency |
| DOPPLER | The Doppler effect (or the Doppler shift) is the change in frequency of a wave (or other periodic event) for an observer moving relative to its source. It is named after the Austrian physicist Christian Doppler, who proposed it in 1842 in Prague. In everyday life the siren of an approaching ambulance will have a higher pitch (frequency) than a siren which has passed by. |
| DF | Direction finding |
| DIFAR | Directional Frequency Analysis and Recording (DIFAR) - A directional passive analogue sonobuoy in service with many countries. Produced by UEL and Sparton |
| DSTL | Defence Science and Technology Laboratory |
| ESE | Experimental active transmit and receive sonobuoy – a UK ship deployed experimental sonobuoy |
| EER/IEER/AEER | Extended(explosive) Echo Ranging/Improved/Advanced – various versions of US multistatic sonobuoys: |
| EER | Impulsive source + DIFAR |
| IEER | Impulsive source + ADAR |
| AEER | Coherent source + ADAR |
| ENIGMA | World Two German Encryption Device used by U-Boats and their command for message security |
| F-size sonobuoy | 4.875 in (123.82mm) diameter, 12 in (304.8mm) long. |
| G-size sonobuoy | 4.875 in (123.82mm) diameter, 16.5 in (419mm) long. |
| GLINT | The higher target echo strength part of a submarine centred on the beam |
| GUIK | that part of the N Atlantic Stretching in a line from Greenland to Iceland to the UK |
| GPS | Hawker Siddeley Aviation |
| HIDAR | High Instantaneous Dynamic Range. The UK's main current passive sonobouy, produced by UEL, and is acoustically equivalent to DIFAR, but with digital sampling, signal processing and telemetry for high dynamic range. |
| HF | High frequency |
| Hydrophone | a device which converts acoustic energy into electrical energy and used in sonobuoys and sonar for detecting sound in the ocean |
| IPT | Integrated Project Team. The term used by UK MOD for the staff managing major defence projects such as the Nimrod MRA4. |

| | |
|---|---|
| JEZEBEL | the name given to the first generation passive narrowband sonobuoys in the 1960s |
| LOFAR | low frequency analysis and ranging explain |
| LOFARGRAM | Hard copy frequency time plot for displaying narrowband tonals emitted from submarines and other ocean based vehicles and objects |
| LF | Low Frequency |
| MSA | Mean Speed of Advance- |
| MSA | Multistatic Active, UEL produced multistatic active search system comprising transmit and receive sonobuoys and airborne processing and displays |
| NDB | Nuclear Depth Bomb |
| MAD | magnetic Anomaly Detection |
| MOD | Ministry of Defence |
| NARROWBAND | Emission or detection of non-coherent acoustic energy over a narrow bandwidth (e.g. acoustic emissions due to rotating machinery) |
| NAVMAR | A US private company which provides ASW expertise |
| Omnidirectional(passive) | A sonar or sonobuoy which receives sounds from all directions |
| Omnidirectional(active) | A sonar or sonobuoy which transmits in all directionsOA - Operational assessment |
| OR | Operational requirement |
| OR | Operational research |
| Orion | the P3 Aircraft |
| QinetiQ | the privatised part of the DERA |
| RADAR | Radio Detection and Ranging. A device for detecting Aircraft developed in the UK in the 1930s |
| RAE | Royal Aircraft (later Aerospace) Establishment |
| Reverberation | Echoes reflected from active transmissions from the sea surface and bottom, bottom features and the sea medium itself. Particularly prevalent in shallow water and can significantly reduce active sonar/sonobuoy detection ranges on submarines. |
| SA | Sea Air |
| SSN | Nuclear powered attack submarine |
| SOSUS | Sound Surveillance System installed by the US during Cold War on the bottom of various oceans for the passive detection of submarines |
| Snorkel | a retractable tube on DESIEL powered submarines for bringing air into submerged submarines and expelling engine exhaust fumes |
| Sea State | A numerical measure of ocean roughness. Usually caused by the prevailing wind |

| | |
|---|---|
| SAR | Search and Rescue |
| SLC | Sonobuoy Launch Container |
| SONAR | SOUND NAVIGATION AND RANGING The name given to ASW passive and active sound detection devices. |
| SR | Staff requirement |
| SNORKEL | A retractable tube on diesel powered submarines for bringing air into submerged submarines and expelling engine exhaust fumes |
| SSN | Nuclear powered attack submarine |
| SSBN | Nuclear powered balllistic missile submarine |
| "super Barra" | (a horizontal planar array, based on the Australian SSQ801 buoy, named 'Barra', which later evolved to become the UK SSQ981E Barra buoy). |
| TTCP | An international organization that collaborates in defence scientific and technical information exchange |
| TONAL | A single narrowband acoustic emission, or related harmonic or related intermodulation component, e.g. due to rotating machinery. |
| VLA | Vertical Line Array of either receive hydrophones or transmit |

# APPENDIX 2

## NATIONAL ARCHIVE SONOBUOY DOCUMENTS

| | | |
|---|---|---|
| **AVIA 18/4089** | *i* Sea Prince T Mk 1 WF.119 (2 Leonides 125): acceptance trials of sonobuoy installation | 1952 Jan 01 - 1952 Dec 31 |
| **DEFE 48/795** | *i* Use of passive directional sonobuoy (barra) screen's in the direct defence of surface forces in the 1980s | 1975 Jan 01- 1975 Dec 31 |
| **DEFE 48/667** | *i* Note on the processing gain of a sonobuoy system operating in Lofar mode | 1973 Jan 01- 1973 Dec 31 |
| **DEFE 48/882** | *i* Evaluation of the relocation and attack capabilities of the Mk 1C active sonobuoy system in the Nimrod aircraft | 1968 Jan 01- 1968 Dec 31 |
| **DEFE 56/77** | *i* The effect of jamming on the employment of airborne Anti-Submarine Warfare Sonobuoy systems | 1972 Jan 01- 1972 Dec 31 |
| **DEFE 56/90** | *i* Accuracy of the Nimrod Sonobuoy D/F and Homer systems at medium level | 1974 Jan 01- 1974 Dec 31 |
| **DEFE 56/120** | *i* Sonobuoy consumption: Nimrod MR Mk 2 | 1975 Nov 01 - 1975 Dec 31 |
| **DEFE 56/135** | *i* Trial instruction: accuracy of the Nimrod sonobuoy homing and 'on-top' indicator systems at medium level | 1973 Jan 01 - 1973 Dec 31 |
| **DEFE 56/140** | *i* Accuracy of the Nimrod sonobuoy homing and 'on-top' indicator systems at medium level. Missing at transfer | 1973 Jan 01 - 1973 Dec 31 |
| **DEFE 67/1** | *i* Evaluation of the active Mk 1c sonobuoy system in squadron service | 1962 Jan 01 - 1962 Dec 31 |
| **ADM 1/24401** | *i* UK/US sonobuoy system: joint Navy/Air Staff requirements for active and passive directional sonobuoys | 1951- 1957 |
| **ADM 1/26861** | *i* Standardisation of existing British and American Sonobuoy systems | 1951 |
| **ADM 1/24436** | *i* Combined Canadian/US/UK sonobuoy system: revision | 1951- 1953 |
| **ADM 1/25319** | *i* Development of low frequency directional sonobuoy system, project `Jezebel' | 1953- 1960 |
| **ADM 219/455** | *i* Anti-submarine defence of convoys: the sonobuoy lane | 1948 |
| **ADM 219/476** | *i* Appreciation of sonobuoy systems | 1951 |
| **ADM 220/278** | *i* VHF switched-cardioid homing device for sonobuoy recovery craft | 1950 |
| **ADM 220/888** | *i* Radio link for ship-projected sonobuoy | 1950 |

| | | |
|---|---|---|
| **AIR 2/12711** | *i* AIRCRAFT: Seaplane and Flying boats (Code B, 5/12): Sunderland aircraft: radio intercommunications, sonobuoy and allied equipment matters | 1938-1950 |
| **AIR 10/6196** | *i* Sonobuoy transmitter type T 7725: general and technical information | 1959-1963 |
| **AIR 10/7776** | *i* ARI 18057: sonobuoy plotting equipment; amendments 1-6 | 1956-1962 |
| **AIR 10/8745** | *i* Sonobuoy Mk 1c receiving and indicating equipment: amendments 2-61 and 64-67 | 1963-1972 |
| **AIR 10/8748** | *i* Sonobuoy transmitter: test equipment; amendments 7-24 | 1963-1967 |
| **AIR 65/244** | *i* Suitability of the Sonobuoy Equipped Raft as a training store Report No: 47/10 | 1947 Sept. 5 |
| **AIR 65/268** | *i* Detection of submarines by aircraft (II): a proposed tactical use of the Passive Directional Radio Sonobuoy Report No: 50/1 | 1950 Mar. 6 |
| **AIR 65/378** | *i* Trial of the British directional sonobuoy Mk 1 series 2: Pt 2B acoustic range of detection of the directional radio sonobuoy Mk 1 series 2 against a medium speed "A" class submarine at depths to 500 feet (trial No 300) | 1953 |
| **AIR 65/379** | *i* Trial of the British directional sonobuoy Mk 1 series 2: Pt 2C maximum ranges of acoustic detection of a fast submarine by the British directional radio sonobuoy (trial No 300) | 1953 |
| **AIR 65/380** | *i* Trial of the British directional sonobuoy Mk 1 series 2: Pt 8 advanced tactical trials (trial No 300) | 1953 |
| **AIR 65/382** | *i* Trials of the sonobuoy attachment ARIX18045 (trial No 328) | 1954 |

| ADM 1/24401 | *i* UK/US sonobuoy system: joint Navy/Air Staff requirements for active and passive directional sonobuoys | 1951-1957 |
|---|---|---|
| ADM 1/26861 | *i* Standardisation of existing British and American Sonobuoy systems | 1951 |
| ADM 1/24436 | *i* Combined Canadian/US/UK sonobuoy system: revision | 1951-1953 |
| ADM 1/25319 | *i* Development of low frequency directional sonobuoy system, project `Jezebel' | 1953-1960 |
| ADM 219/455 | *i* Anti-submarine defence of convoys: the sonobuoy lane | 1948 |
| ADM 219/476 | *i* Appreciation of sonobuoy systems | 1951 |
| ADM 220/278 | *i* VHF switched-cardioid homing device for sonobuoy recovery craft | 1950 |
| ADM 220/888 | *i* Radio link for ship-projected sonobuoy | 1950 |
| ADM 1/24401 | *i* UK/US sonobuoy system: joint Navy/Air Staff requirements for active and passive directional sonobuoys | 1951-1957 |
| ADM 1/24525 | *i* Sonobuoys: report on operating experiences with type 1946 | 1953-1954 |
| ADM 219/471 | *i* Endurance required of sonobuoys | 1951 |
| ADM 219/659 | *i* The use of RAP sonobuoys for datum search | 1973 |
| ADM 219/678 | *i* A study into the operational utilisation and performance of Doppler RAP sonobuoys | 1974 |
| ADM 219/716 | *i* War reserves of the Sea King Replacement's (SKR's) sonobuoys | 1978 Apr 01 - 1978 Apr 30 |
| ADM 204/3478 | *i* A doppler tracking algorithm for use with non-directional passive sonobuoys | 1975 Dec 01 - 1975 Dec 31 |
| ADM 204/3518 | *i* Target location from passive sonobuoys using a Doppler method | 1976 Nov 01 - 1976 Nov 30 |
| ADM 253/620 | *i* Selection of material for shock testing of compasses in sonobuoys | 1951 |
| ADM 259/205 | *i* LOFAR techniques applied to sonobuoys | 1953 |

| AVIA 89/249 | *i* Service approval testing of sonobuoy type T.24501 | 1969 Jan 01 - 1969 Dec 31 |
|---|---|---|
| AVIA 89/294 | *i* Interim (31 day) report on design approval tests on contractor demonstration models of sonobuoy T 17164 (contract K19B/133/CB/LR/12(b)). With photographs | 1973 Jan 01 - 1973 Dec 31 |
| AVIA 89/296 | *i* Design approval tests on contractor demonstration models of sonobuoy T 17164. With photographs | 1973 Jan 01 - 1973 Dec 31 |
| AVIA 89/306 | *i* Design approval tests on contractor demonstration models of sonobuoy T 17164 (second submission). With photographs | 1975 Jan 01 - 1975 Dec 31 |
| AVIA 89/307 | *i* Addendum to EQD Report No 1/75 on design approval tests on contractor demonstration models of sonobuoy T 17164 (second submission) | 1975 Jan 01 - 1975 Dec 31 |
| AVIA 6/17649 | *i* Possible methods of achieving a long range passive sonobuoy system | 1958 |
| AVIA 6/17652 | *i* Assessment of active directional sonobuoy performance | 1959 |
| AVIA 6/17666 | *i* Directional explosive echo ranging sonobuoy investigation | 1963 |
| AVIA 6/17668 | *i* Directional explosive echo ranging sonobuoy trials | 1964 |
| AVIA 6/19486 | *i* Mechanical strength tests on British directional sonobuoy X1908 | 1955 |
| AVIA 6/19998 | *i* Assessment of a directional explosive echo ranging sonobuoy system | 1966 |
| AVIA 6/20437 | *i* Parachute assembly for the sonobuoy transmitter T1939 | 1952 |
| AVIA 6/20971 | *i* Interference to the sonobuoy Mk 1C system from mobile communications equipment | 1964 |
| AVIA 6/25077 | *i* Sonobuoy development programme: UHF propagation experiments over paths close to the sea surface | 1975 Jan 01 - 1975 Dec 31 |
| AVIA 6/25376 | *i* Short study of detection and tracking performances of directional passive sonobuoy systems and Nimrod operations against high-speed submarines | 1976 Jan 01 - 1976 Dec 31 |
| AVIA 18/3942 | *i* Shackleton MR Mk 2 WG.530 (4 Griffon 57): acceptance trials of Avro sonobuoy carriers, six store, front and rear | 1953 Jan 01 - 1953 Dec 31 |
| AVIA 18/4078 | *i* Sea Prince T Mk 1 G-ALCM: sonobuoy installation in Sea Prince T Mk 1 | 1951 Jan 01 - 1951 Dec 31 |

# APPENDIX 3

## A US VIEW OF THE FUTURE OF AIRBORNE ASW

Reproduced by kind permission of Peter Verburgt.

Where might technology take Air ASW in the future is difficult to predict, but there are a few clues. The clues are not just technology based however. Computer chips will undoubtedly operate faster, integrated circuits will likely become smaller, but they alone will not drive the future direction for the Navy. I submit that the clues or roadmap to the future are embedded in cost, threat, environment, aircraft platforms, exploitables, as well as those breakthroughs enabled by technology advances. To expand just a bit on these:

a. **"Cost"** — This is an ever-important reality of the landscape today. Techniques to reduce system cost such as reducing reliance on expendables will be highlighted over those that are costlier.

b. **"Threat"** — There is somewhere over five hundred known submarines in the world today with over one hundred more in the planning stage. A number of these threats are very serious in their stealth and weapon delivery ability. Our ASW developments will be stressed by the improvements required to detect these threats.

c. **"Environment"** — Deep environments have a different ASW solution set from shallow littoral areas, and that includes not just the available acoustic propagation paths but the RP environment in which our sensor systems must operate. Denied area environment are yet another complication in search of a solution.

d. **"Aircraft Platforms"** —The Navy 's newest platform is the P-8 Poseidon aircraft. It has a new expanded flight envelope over its predecessor. This offers new opportunities for ASW sensor systems to expand capabilities in synergy with the new platform. Beyond are opportunities for the burgeoning field autonomous aircraft (large and small sonobuoy size), and even beyond is the potential for high altitude airships and space based. Each platform exposes new technology potential.

e. **"Exploitables"** — Much is known today about acoustics and its application to ASW. As the threat submarine evolves, new exploitable signatures are likely to be generated. These signatures may or may not be detectable acoustically, so that added attention will need to be paid to alternate non-acoustic technology.

f. **"Technology Advances"** — Not to be overlooked, technology itself will continue to follow a progression of advances. Cheaper and smaller breakthroughs will lead to new generation of sensors with expanded ASW applications and perhaps capabilities.

The future is wide open in the area of Airborne ASW. The clues are there for directing the research, and platforms are awaiting enhanced capabilities.